Annual Meeting of the American Electrophoresis Society 2016 (AES)

Topical Conference at the 2016 AIChE Annual Meeting

San Francisco, California, USA
13 - 18 November 2016

ISBN: 978-1-5108-3419-4

Printed from e-media with permission by:

Curran Associates, Inc.
57 Morehouse Lane
Red Hook, NY 12571

CURRAN ASSOCIATES INC. proceedings .com

Some format issues inherent in the e-media version may also appear in this print version.

Printed by Curran Associates, Inc. (2017)

AIChE
120 Wall Street, FL 23
New York, NY 10005-4020

Phone: (800) 242-4363
Fax: (203) 775-5177

www.aiche.org

Additional copies of this publication are available from:

Curran Associates, Inc.
57 Morehouse Lane
Red Hook, NY 12571 USA
Phone: 845-758-0400
Fax: 845-758-2633
Email: curran@proceedings.com
Web: www.proceedings.com

TABLE OF CONTENTS

463152 *In Vitro* Electrical Impedance Characterization of Huvecs Undergoing Hydrodynamic Shear Stress

Abstract:

The pathophysiological processes of atherosclerosis [1], [2], thrombosis [3], [4], and their related complications like cardiovascular diseases have all been linked to increased endothelium permeability. Though studies have strived to understand endothelial biology and kinetics through *in vitro* models, the majority of endothelium studies have been primarily with static cell culture models or macro-scale models that rely on end-point analysis, as well as, do not properly represent the physiological sizes, structures, and environmental conditions of human blood vessels. This work presents a microfluidic platform that replicates physiological shear stress/fluid flow conditions and the length scale of the blood vessels with the ability to electrically characterize the endothelium morphology and permeability. An electrical impedance biosensor provides a real time evaluation of cell-cell interaction and cell migration [5]. The microfluidic chip consists of two arrays of independently addressable gold electrodes of various diameters (50 μm, 100 μm, and 200 μm), and a common large counter electrode. Different electrodes sizes were used to test and compare the behavior of different subpopulations sizes within the monolayer. The microfluidic chip also includes a two-channel design, such that one may contain cells, while the other remains without cells. This design feature helps to identify environmental variables affecting the endothelial cell data. Human umbilical vein endothelial cells (HUVECs) were used as an initial endothelium model and were cultured on top of micro-patterned gold electrodes. A peristaltic pump was connected to the chip subjecting the endothelial cells to a constant shear stress conditions (15 dyne/cm^2), a stepped shear condition of (20-30-50 dyne/cm^2), and a control static (0 dyne/cm^2) condition. An in-house designed printed circuit board was programmed to automatically record impedance spectra from a high-precision impedance analyzer (HP4294a) at specified time intervals. Impedance spectra (100 Hz- 1 MHz) collected were fitted to an equivalent circuit model in order to extract the cell behavior parameters such as cell monolayer resistance (R_{TER}). Experimental results show there is a shear-dependent increase in cell monolayer resistance at the onset of shear and it is greater for higher shear stress values. The cell monolayer resistance jump is only observed for acute flow as sheared cells do not generate a substantial increase in cell monolayer resistance when subjected to a higher shear stress. For both the constant shear condition and stepped shear condition, this increase is temporary and proceeded by a decline before it begins to stabilize below the baseline value (measurement at the onset of shear). Although the work presented here, focuses on general endothelium studies under fluid flow, this platform is very versatile. Other possible applications include drug testing, brain barrier characterization, role of mechanical stimuli in tumor metastasis and wound healing.

470244 Utilization of Direct Current Insulator-Based Dielectrophoresis in the Separation of Breast Cancer Infected Peripheral Blood Mononuclear Cells from Their Mixture with Healthy Cells

Abstract:

Currently, early stage breast cancer detection is a major challenge faced by developed and developing countries. Detecting this diseased cells at the teething stage can forestall the development of diverse complications including metastasis. In this work, a numerical simulation was used to characterize and separate breast cancer infected peripheral blood mononuclear cells (iPBMCs) from their mixture with healthy mononuclear cells (hPBMCs) via direct current insulator-based dielectrophoresis (DC-iDEP) within a microfluidic platform. COMSOL Multiphysics v5.2 was used to simulate the device geometry consisting of a microchannel incorporating semicircular obstacles and bifurcating into two outlet channels with different applied voltages and different inlet concentrations in the range of 1 iPBMC per 10 - 10^8 hPBMCs. Current conservation along with fluid flow, convection, and diffusion equations were used to predict the concentration distribution of the cells. Results indicated that the microdevice (which had a series of four 0.07 mm diameter semi-circular obstacles) when operated at an inlet electrode voltage of 130 V, and 0/50V at the outlet channels was able to separate iPBMCs from hPBMCs irrespective of the inlet percentage composition of the iPBMCs. The device has been fabricated and experimental validation of the simulation results is ongoing. The results obtained from the experiment will be compared with the simulationâ€™s and relevant analyses will be made.

467969 3D Electrodes Integrated in Microfluidic Channels for Automated Single Cell Electrorotation Spectra Acquisition

Abstract:

3D electrodes integrated in microfluidic channels for automated single cell spectra acquisition

Samuel C. Kilchenmann, Iness R. Benmessaoud, Pietro Maoddi, Marta A. Comino, Carlotta Guiducci

Electrorotation is a known technique to extract electrical parameters of cells. As the rotational torque is highly dependent on the electric field strength, previous works focused on optimizing the shape of planar electrodes, to achieve uniform field strengths. Mostly, the optimization was performed only in 2D space and electrode distances above 100 µm were considered. While this strategy might lead to valid results in the electrode center, it is not suitable in cases of small electrodes, where the inter-electrode distance is in the range of the channel height, as the electrical fields will have a strong curvature and dependence on height. For this reason, in our lab, we focus on the application of 3D electrodes for electrorotation experiments. Such electrodes allow to achieve electric fields that have no dependence on the z-position and thus the rotation torque is constant over the full channel height. A direct consequence is that the signal amplitudes can be lowered, as for a given signal amplitude, 3D electrodes lead to a higher torque moment inside the microfluidic chamber, compared to a similar planar design. Ultimately 3D electrodes reduce the risk of cell and electrode damage, as they lead to lower electric field peaks and to lower thermal heating effects.

Furthermore, 3D electrodes were shown to lead to more stable particle trapping in microfluidic channels [1]. This can be used to achieve simultaneous trapping and rotation of particles inside microfluidic channels by applying simultaneous DEP and ROT signals [2]. For this reason, we developed a fabrication process to achieve 3D electrodes integrated in microfluidic channels [3] and we currently employ electrorotation experiments to extract complete electrorotation spectra of one single cells. The cells are injected into the system by microfluidic sample handling with a pressure driven system that allows for high control on flow and instantaneous flow stop. The latter in combination with the DEP trapping, allows for high precision in cell positioning. The good control over the electric field and the precise positioning of single cells by means of DEP lead to high control of experimental conditions and side effects such as clustering and particle-particle interactions can be reduced. For this reason, this system allows to extract a single cell electrorotation spectrum and to compare it to the rest of the population.

[1] J. Voldman, M. Toner, M. L. Gray, and M. A. Schmidt, "Design and analysis of extruded quadrupolar dielectrophoretic traps," J. Electrostat., vol. 57, no. 1, pp. 69–90, 2003.
[2] A. Rohani, W. Varhue, Y.-H. Su, and N. S. Swami, "Electrical tweezer for highly parallelized electrorotation measurements over a wide frequency bandwidth.," Electrophoresis, vol. 35, no. 12–13, pp. 1795–802, Jul. 2014.
[3] S. C. Kilchenmann, E. Rollo, P. Maoddi, and C. Guiducci, "Metal-Coated SU-8 Structures for High-Density 3-D Microelectrode Arrays," J. Microelectromechanical Syst., pp. 1–7, 2016.

469362 Quantifying Intracellular Mitochondrial Dynamics Based on Cytoplasmic Electrophysiology

Abstract:

Mitochondria, which serve as energy centers within the cell, undergo constant fusion and fission, thereby allowing the cell to quickly adapt to environmental conditions for promoting mitochondrial health. Mutation and aberrant regulation of the mitochondrial machinery is associated with a number of human diseases including Parkinson's Disease, Alzheimer's Disease and Diabetes, as well as physiological processes whose dysregulation are classical hallmarks of human cancer. In recent work, we demonstrated that the activation of mitochondrial fission is required for tumor growth in a xenograft model of pancreatic cancer and inhibition of mitochondrial fission was shown to prevent tumor growth [1]. However, therapies targeting aberrant regulation of mitochondrial machinery require label-free means to quantify the dynamic alterations in mitochondrial morphology after particular genetic screens as well as towards enabling the collection live cells with a particular mitochondrial structure for downstream analysis to understand the signaling pathways. Herein, utilizing two different cell types, namely mouse embryonic fibroblasts and human embryonic kidney cells to create modified mitochondrial phenotypes by entirely independent means, we show that variations in the intracellular cytoplasmic conductivity can serve as a means for quantifying alterations in mitochondrial structure after particular genetic screens. This difference in cytoplasmic conductivity is shown to be capable of enabling frequency-selective dielectrophoretic isolation of cells with a particular mitochondrial structure. Based on this, we envision a label-free platform to aid the discovery process for development of therapeutics targeting the mitochondrial machinery.

[1] Kashatus, J.A., Nascimento A., Myers, L.J., Sher A., Byrne F.L., Hoehn K.L., Counter C.M., and Kashatus D.F. Erk2 Phosphorylation of Drp1 Promotes Mitochondrial Fission and MAPK-Driven Tumor Growth. Mol. Cell 57, 537–551 (2015)

468430 Integrating Micro-Environmental Cues into Single-Cell Targeted Proteomics Tools

Abstract:

Variation among individual cells does not arise in isolation, with the local microenvironment understood to play an important role. Physiologic processes underpinning stem cell differentiation and maturation are impacted by a complex milieu of extracellular factors including paracrine signaling, proximity and characteristics of neighboring cells, and mechanical substrate properties. Direct measurement of protein-mediated signaling in individual cells sheds insight on cellular response to each environmental variable, including 3D culture. Concomitantly, our previous studies show that microfluidic electrophoresis deepens the protein profiling that is possible in cytometry (analysis of single cells). Here, we advance the design of single-cell resolution electrophoretic separations to now include analysis of cells that are not in isolation.

To accomplish this goal, we design, demonstrate, and characterize a microfluidic cell patterning technique that allows us to culture single cells and small cell clusters in controlled, tunable microenvironments and then directly analyze the lysate of each single cell *in situ*. Using short assay times, electrophoresis, and photo-activatable crosslinkers, we demonstrate separation, detection, and ratiometric analysis of protein targets from stem cells in diverse microenvironments. We describe our multi-layered (3D) microfluidic device, tunable polymer formulations (chemical, mechanical, geometric) and separations-related outcomes (dispersion, dilution, separation resolution, peak capacity). By integrating microenvironmental cues into a precision microanalytical instrument, we measure key morphometric and spatial information and correlate these physical data to protein expression patterns of each mammalian cell.

Looking forward, we view the combination of spatio-temporal information with targeted proteomic information as an important and, as of yet missing, axis of systems biology.

472390 Dielectrophoretic Field Shaping for Enhanced Circulating Tumor Cell Isolation and Characterization

Abstract:

Isolation or enrichment of viable cell subpopulations from a complex biological sample is of paramount importance in many biomedical applications requiring quantification or the molecular characterization of a particular cell fraction. As of recent, dielectrophoresis (DEP) has attracted various research groups for its potential as a sample enrichment method. Most often varied complications are suffered when traditional devices use DEP as an enrichment method. Complications include low throughput, electrode and sample fouling, bubble formation, and undesired electrochemical reactions that could damage the array of electrodes used. These problems have limited DEP device utilization in practical biomedical applications. The device presented in this work possesses simple new manufacturing techniques that eliminates all the mentioned complications while reducing sample processing time, reduces prototyping cost and shrinks the gap to mass production. The DEP device proposed in this work creates a quasi-constant dielectrophoretic force along entire length of the microfluidic channel, enhancing cell separation. This DEP device isolates cultured cancer cell lines from peripheral blood mononuclear cells (PBMC). SK-BR-3, MCF-7 (human breast adenocarcinoma) and PC-3 (human prostate adenocarcinoma) are cancer cell lines used in this study that capture efficiencies in the order of 80, 82, and 86%. Respectively, PBMC depletion rates ranging from 90-95% are present.

471312 Separation of *Candida* Cells Using 3D Carbon-Electrode Dielectrophoresis

Abstract:

Separation of *Candida* Cells using 3D Carbon-electrode Dielectrophoresis

Jordon Gilmore, Monsur Islam and Rodrigo Martinez-Duarte

Mechanical Engineering Department, Clemson University, Clemson, SC, USA

Here we present preliminary results towards separation of *Candida* cells using 3D carbon-electrode dielectrophoresis (DEP). In particular, two *Candida* species have become the most isolated in instances of systemic candidiasis, accounting for nearly 75% of all cases.[1] These species, *Candida albicans* and *Candida glabrata* are usually found in the oral cavity and intestinal tract of healthy humans. However, given certain conditions, both of these species may become pathogenic.

The goal of this work is to employ DEP separation techniques with carbon electrodes to isolate *C. albicans* and *C. glabrata* from a heterogeneous solution comparable to a biological sample from the clinical setting. The authors demonstrate successful trapping of each species by positive dielectrophoresis (pDEP). This is a preliminary step in the goal to separate *Candida* cells by species as part of a point-of-care diagnostic tool for determining the dominant *Candida* species during suspected candidiasis. This is important given the discovery that all *Candida* species are not responsive to the same treatments (i.e. *C. glabrata*, which do not respond to the common fluconazole treatments of *C. albicans* – based candidiasis [2]).

Strains of *C. albicans* and *C. glabrata* were cultured in Yeast Malt Broth (YMB) supplemented with a 40% glucose solution at a 1:100 glucose to YMB ratio. The YMB media was prepared and acidified to a pH of 3.5 by the addition of Citric Acid. Cells were grown in suspension and dynamic incubation (shaking at 215 rpm) at 28°C for two days. Average diameters of *C. albicans* and *C. glabrata* were 4.68 ± 0.97 µm and 3.01 ± 0.49 µm, respectively.

For DEP experiments, the experimental media used for cell suspension and buffer solution was a 15% sucrose, 0.3% dextrose, and 0.1% bovine serum albumin (BSA) mixture in DI water. Conductivity of the media was 12.62 µS/cm. To prepare the experimental cell suspension, 150 µl of the cell culture was suspended in 4 ml of the experimental media followed by washing and re-suspending in experimental media. The cell concentrations of the cell suspensions were 7.18×10^6 cells/ml and 1.92×10^7 cells/ml for *C. albicans* and *C. glabrata* respectively.

The fabrication of the carbon-electrode DEP device has been detailed by our group in previous publications[3]. The cells were flowed through the DEP chip at a constant flow rate of 2.5 µl/min using a syringe pump. The electrodes were polarized with a sinusoidal signal with constant magnitude of 20 V_{pp} and varying frequencies ranging from 10 kHz to 5 MHz. The effect of the frequencies on the pDEP has been investigated for both species using an optical microscope. The qualitative results based on the microscopic observation were plotted in Figure 1. For *C. albicans* the cross over frequency is in between 750 kHz to 1 MHz, whereas the cross over frequency for *C. glabrata* is in between 2.5 MHz and 5 MHz. Also Figure 1 shows the difference in the cell trapping over the change of the frequencies. Figure 2 shows the DEP trapping of the cells at 100 kHz frequency.

As the size of *C. albicans* and *C. glabrata* are close and they belong to same species, their DEP response is quite similar. Therefore, the ongoing work focuses on determining the specific DEP parameters that allow for the separation of each species. Separation of these species will allow for identification of the primary cell type responsible for the candidiasis case, allowing physicians to provide more optimized treatments.

Acknowledgements

The authors acknowledge Dr. Mark Blenner's research group from Chemical Engineering Department for assistance with the cell culture.

References

(1) Chakravarthi, S.; Haleagrahara, N. A Comprehensive Review of the Occurance and Management of Systematic Candidiasis as an Opportunistic Infection. *Microbiol. J.* **2011**, *1* (1), 1–7.

(2) Hitchcock, C. A.; Pye, G. W.; Troke, P. F.; Johnson, E. M.; Warnock, D. W. Fluconazole Resistance in Candida Glabrata. *Antimicrob. Agents Chemother.* **1993**, *37* (9), 1962–1965.

(3) Martinez-Duarte, R.; Renaud, P.; Madou, M. J. A Novel Approach to Dielectrophoresis Using Carbon Electrodes. *Electrophoresis* **2011**, *32* (17), 2385–2392.

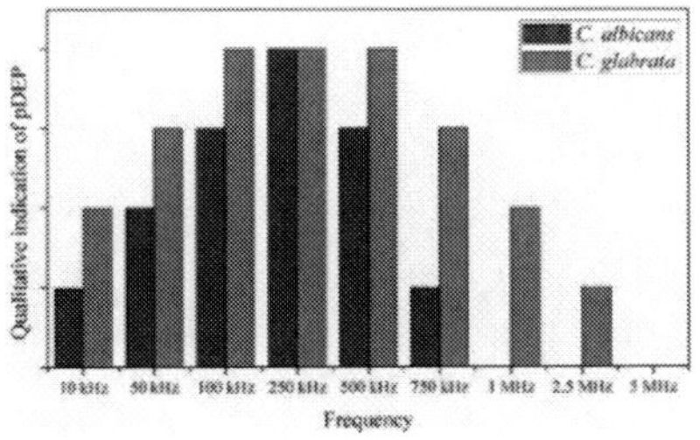

Figure 1: Qualitative indication of positive DEP (pDEP) under microscopic observation for *C. albicans* and *C. glabrata* for different frequencies

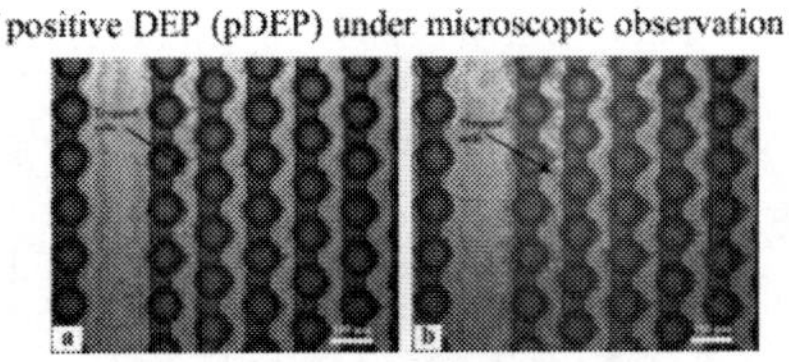

Figure 2: Trapping of (a) *C. albicans* and (b) *C. glabrata* cells around the carbon electrodes. The DEP parameters were 100 kHz frequency and 20 V magnitude. The images were captured 2 minutes after the electric field was turned on.

470835 High-Throughput, Low-Loss Dielectrophoretic Cell Separation

Abstract:

Worldwide, the market for cell separation is dominated by three methods; fluorescence-activated cell sorting, magnetic-activated cell sorting, and density gradient methods; these variously suffer from drawbacks including high cell loss (typically >50%), the requirement for fluorescent or magnetic labels, and high cost (capital, operating or both). Dielectrophoresis was first used for cell separation over 50 years ago, and numerous systems for dielectrophoretically-activated cell sorting have been presented but few have matched existing methods for throughput and recovery. We have developed a cell enrichment system using the DEP-Well platform. Cells pass through 343 parallel wells containing 12 electrodes apiece, and energised by a 14Vpp signal at frequencies up to 1MHz. Processing speed is sufficiently high (in excess of 1 million cells per ml), and cell losses sufficiently low (below 5%), to enable multiple passes through the device in order to improve performance; on a second pass, separation efficiencies in excess of 95% are readily achievable; whilst on a third pass, recovery of rare cells becomes possible. Overall processing speed and recovery are comparable to or exceed FACS and MACS, whilst the use of a disposable chip avoids inter-separation contamination.

469750 Dielectrophoretic Sorting of Plasmid and Genomic DNA

Abstract:

Innovations in DNA sample preparation may soon enable development of rapid sequencing and bioanalysis tools that significantly improve the availability and accuracy of nucleic acid analysis. Over the years, a variety of methods have been used to separate nucleic acid fragments based on size. Since various Next-Generation Sequencing (NGS) methods require DNA fragments of specific size, the need for improved sorting has fueled significant analytical research. Achieving efficient and continuous flow fractionation of DNA and RNA still presents considerable challenges. Current separation methods have not fully satisfied the need for rapid fractionation of heterogeneous nucleic acid samples into populations of specific and narrow size distribution. We have recently explored the use of dielectrophoresis to improve sized based, continuous sorting with reduced analysis times. For this purpose, we have developed a microfluidic constriction-sorter device, using insulator-based dielectrophoresis and pressure-driven flow. This approach enables continuous-flow separation of nucleic acid analytes into physically distinct microchannel outlets. We demonstrate AC dielectrophoretic sorting of four different DNA analytes, including both plasmid and genomic DNA. The sorting behavior is tunable and size-specific. It demonstrates strong dependency on applied potentials and frequencies. Specifically, we have investigated the behavior of 1.0, 10.2, 19.5, and 49.5 kbp dsDNA. Applied potentials ranged from 200 to 2400 V/cm, while frequencies ranged from 50 Hz to 20 kHz with flow rates in the range of 1 μL/h. These conditions yielded a maximum observed sorting efficiency of approximately 92%. Based on these results, we propose the use of a continuous-flow fractionation strategy to facilitate future coupling with NGS approaches.

459207 Alternating Current Electro-Osmotic Pumping at Asymmetrically Metallized Porous Membranes

Abstract:

Alternating current electro-osmotic pumping at asymmetrically metallized porous membranes

J. Beharic and C. K. Harnett

Department of Electrical and Computer Engineering, University of Louisville

The goal of this project is to investigate AC electro-osmotic pumping through asymmetrically metallized membrane nanopores for bubble-free, electrically driven fluid transport in compact lab-on-a-chip systems that require higher backpressure than can be achieved with microscale pumping features [1]. Induced Charge Electroosmosis (ICEO) is an electrokinetic phenomenon at the interface of polarized conductive surfaces and electrolytes in the presence of an electric field. The polarized conductor attracts counterions, forming a double layer at the surface. With continuous application of an electric field, the ions in the double layer move with a slip velocity that can create bulk flow. ICEO flow can persist even in alternating current (AC) electric fields [2]. Most work in this area uses microfabricated pillars, in contrast to the flow-through format described here. We have created several different types of metallized membranes and tested them under different conditions. To validate our experimental results and to optimize our parameters we have also performed computational fluid dynamics (CFD) simulations. The work is motivated by the idea that nanoscale pore geometry can potentially exceed the pressure limitations of current ICEO and alternating current electroosmosis (ACEO) microscale systems. Nanoporous materials made either by self-assembly, or by "top-down" nanolithography, will open the possibility of using ICEO in new applications including lab-on-a-chip sample preparation.

Methods

Fabrication

We have produced flow-through ICEO structures using both top-down photolithography for optical microscopy experiments, and track-etched nanoporous polymer membranes for pressure measurements. Unlike traditional micropillar-based ICEO devices, the fabrication of these asymmetrically metallized membranes is a simple process. Nanoporous ICEO structures are fabricated by sputtering gold onto off-the-shelf track-etched nanoporous polycarbonate membranes. The sputtering process coats only the top and side walls of the membrane, breaking the symmetry of the metal coating. Track-etched pores are commercially available at nanoscale diameters (80 nm- 1 micron) which are not achievable using photolithography. Although the track-etched membranes provide nanoscale features for backpressure generation, their random distribution and small size makes it difficult to optically image the flow pattern at individual pores. Therefore, larger and more uniform pores are needed for optical studies. These larger pores were created using SU-8 epoxy-based photoresist and standard cleanroom lithography processes. The membranes were created by first depositing a sacrificial layer of styrene onto a wafer, followed by standard SU-8 lithography and release in toluene. The released SU-8 membranes were transferred to a silicone carrier and metallized by sputtering.

Measurements

Pressure measurements were performed with a MPXV7002DP pressure sensor which outputs a voltage proportional to pressure difference across the membrane. The voltage was collected using LabVIEW and analyzed in MATLAB. The pressure measurement device consisted of two reservoirs connected by a microfluidic channel. Electrical contact was made using platinum wires immersed in the reservoirs. The membrane was placed above one end of the connecting channel. By using an indium tin oxide slide as a transparent electrode and substrate for the microfluidic channel, the membrane could be polarized while tracer particles were observed using inverted fluorescence microscopy. The flow was recorded using an INFINITY3-3ur video camera and analyzed using MATLAB Open PIV.

Simulations

Simulations were performed in ANSYS Fluent CFD software using a moving wall in an axisymmetric domain, representing electroosmotic slip along the pore walls. The surface charge density along the wall was calculated using ANSYS Maxwell. The resulting charge density distribution was then imported into the ANSYS Fluent mode to calculate flow velocity vectors. The goal was to determine the net flow, which combines ICEO-driven flow with pressure-driven backflow, and the maximum pressure and flow rates attained by a given geometry. Maximum velocity was calculated by leaving the inlet and outlet open. Using these simulations allows us to determine the optimal parameters to optimize our device performance.

Results

Out-of-plane tracer particle motions were observed during fluorescence microscopy of active pores in AC electric fields. Pressure measurements depended on several variables:

- Voltage. Increasing the voltage (0-50Vpp) resulted in an increase in pressure, with typical maximum pressures in the 120 Pa range for a single membrane and higher for stacked membranes.
- Frequency. Higher driving frequencies (> 1000 Hz) quenched the ICEO effect as the ions in the fluid could not diffuse fast enough to respond to charges on the metal surface.
- Electrolyte conductivity. Increasing KCl concentration in the solution diminished the ICEO effect. The drop-off in ICEO slip velocity with increased conductivity is consistent with the literature, and is usually ascribed to ion crowding [1,4].
- Pore diameter. The pressure output of the devices increased with decreasing pore diameters as predicted by CFD simulations. Optimal performance was found at 400 nm.

Conclusions

We have successfully created metallzed membranes for induced charge electroosmotic flow. Flow has been validated by fluorescence video microscopy at membranes using fluorescent tracer beads. The flow response to the AC driving signal has been validated by pressure readings. Pressure ranges are comparable to most conventional ICEO devices, with simpler fabrication.

[1] C. C. Huang, M. Z. Bazant and T. Thorsen, Ultrafast high- pressure AC electro-osmotic pumps for portable biomedical microfluidics, Lab Chip, 2009, 10(1), 80–85.

[2] Bazant, M. Z., & Squires, T. M. (2004). Induced-charge electrokinetic phenomena: theory and microfluidic applications. *Physical Review Letters*,*92*(6), 066101.

[3] C. K.Harnett, Jeremy Templeton, Katherine A.Dunphy-Guzman, Yehya M.Senousy, and Michael P.Kanouff, Lab on a Chip, 8 , Pages 565-572, 2008

[4] Bazant, M. Z., Kilic, M. S., Storey, B. D., & Ajdari, A. (2009). Towards an understanding of induced-charge electrokinetics at large applied voltages in concentrated solutions. Advances in Colloid and Interface Science, 152(1-2), 48-88.

464737 Rapid Cell Separation Using 3D Carbon Electrode Dielectrophoresis

Abstract:

Rapid cell separation using 3D carbon electrode dielectrophoresis

Monsur Islam, Rucha Natu and Rodrigo Martinez-Duarte

Mechanical Engineering Department, Clemson University, Clemson, SC, USA

Here we present the initial results for rapid separation of two different particles from a particle mixture using carbon electrode dielectrophoresis (DEP). Isolation and concentration of pathogenic cells are one of the important steps in clinical diagnostic systems. Expensive and large laboratory equipment such as fluorescence activated cell sorting (FACS) are generally used for cell isolation; in a process that is relatively expensive and can take several hours. Here we propose a DEP-based microfluidic platform for rapid separation and concentration of cells. Preliminary results indicate separation of *Candida albicans* and latex beads of 10 µm diameter from the cell-particle mixture at a flow rate of 100 µl/min using streamingDEP.

C. albicans cells were grown in yeast malt broth (YMB, Sigma Aldrich, USA) culture for 2 days in a controlled environment at 37 °C. 150 µl of this cell culture was then inoculated in 4 ml of experimental buffer solution which contained 15% sucrose, 0.5% dextrose and 0.1% bovine serum albumin (BSA). The experimental cell suspension was prepared by washing and re-suspending the cells in the buffer solution. The concentration of the cell suspension was 7.18×10^6 cells/ml. A particle suspension of 10 µm latex beads was also prepared in the buffer solution and concentration was measured 5.86×10^6 particles/ml. A mixture solution was also prepared by mixing the cell suspension to the particle suspension in a 1:1 volume ratio.

The fabrication of the carbon electrode DEP device was reported elsewhere by our group [1]. DEP experiments were performed for both cell suspension, particle suspension and mixture solution. The carbon electrodes were stimulated with sinusoidal signal having magnitude of 20 Vpp and frequency of 250 kHz. At this conditions, *C. albicans* experience positive DEP and the 10 µm latex beads experience negative DEP. A constant flow rate of 100 µl/min was maintained throughout the experiments. At the cell density used in this work this would represent a throughput of $\sim 10^5$ cells/min. The continuous separation principle shown here is based on attracting the targeted cell towards the electrode but not trapping it. Hence, an equilibrium between the positiveDEP and hydrodynamic drag force is sought. Streaming behaviour of *C. albicans* is illustrated in Figure 1a. Instead of streaming, latex beads experience focusing by negativeDEP and were focused in between the electrodes (Fig. 1b). When using the mixture solution, both the above mentioned phenomena occurred simultaneously and *C. albicans* and latex beads focused and flowed along different streamlines showing separation in Fig 1c. The results obtained when the electric field is off are shown in Fig. 1d, showing how the lines disappeared.

Ongoing work is to determine the maximum flow rate while still observing streaming behaviour. To this end, COMSOL is being used to determine the position of the streamlines at different flow rates. The fabrication of geometries that would allow for the continuous retrieval of the focused cells represents future efforts in this work.

References:

[1] R. Martinez-Duarte, P. Renaud, and M. J. Madou, "A novel approach to dielectrophoresis using carbon electrodes.," *Electrophoresis*, vol. 32, no. 17, pp. 2385–92, Sep. 2011.

Figure 1: Focusing of (a) *C. albicans* using positive DEP, (b) 10 µm latex beads using negative DEP; (c) Separation of *C. albicans* from 10 µm latex beads; (d) No focusing is observed when the field is turned off. The DEP parameters were 20 Vpp magnitude and 250 kHz frequency and the flow rate was 100 µl/min. Images taken at a rate of 100 fps.

474650 New Diagnostic Paradigms Enabled By Hyperresolution Dielectrophoretic Separations

Abstract:

Our group is developing better separation methods for targets ranging from proteins to ten micron particles, including virtually all biological targets (proteins, organelles, bioparticles, cells, etc.) Using gradient techniques for separations of complex mixtures allows for the effects diffusion and other dispersive factors to be minimized. An especially attractive mix of forces for common biological samples is electrophoresis and dielectrophoresis. We have a closely related approach term electrophoretic exclusion which includes a flow field, which I will cover briefly, but the approach I will mostly focus on is one that sets electrophoretic forces to oppose dielectrophoretic ones across an open channel. In this channel, there are unique capture points for various values of electrophoretic and dielectrophoretic properties of the analytical targets, described by characteristic mobilities (u_{EK} and u_{DEP}, respectively). With this technique, very similar particles and biological materials can be efficiently separated. Under optimal conditions a one micron particle can be separated from one that varies by only one nanometer (one part in a 1000-2000). Other characteristic resolution factors include minimal resolvable u_{DEP} at 10^{-26} m^4/V^2s (one part in 10^9) and minimal resolvable Clausius-Mossotti factor at 10^{-9}. These are extraordinary figures of merit. We have already shown extremely positive results by separating gentamicin resistant *Staphylococcus epidermidis* and *Staphylococcus aureus* from susceptible strains, reflecting extremely subtle alterations in cellular structure or function. A theoretical framework and resulting predictions will be presented along with key data supporting the models.

468278 Separation of Hemocrit from Hemolymph Collected from Individual Drosophila

Abstract:

Although Drosophila melanogaster is an important animal model for many human diseases, the small size of Drosophila typically limits the volume of sample that can be collected which complicates obtaining chemical content data. Work in our lab has focused on the collection and chemical analysis of the tens of nanoliter volumes of hemolymph that can be collected from individual Drosophila. The blood-like hemolymph plays roles of chemical delivery and waste removal and also contains hemocyte cells that provide an immune response. All previous work has employed whole-hemolymph, plasma and hemocyte content, for chemical analysis including quantitation of amino acids and proteomics. The goal in this work is to separate the cellular content from the hemolymph plasma for subsequent analysis. Following collection of hemolymph from individual Drosophila, hemolymph is diluted to slow the clotting reaction. A number of strategies were tested for separation of hemocrit including filtering schemes and centrifugation. Hemocyte content was greatly reduced by centrifugation by not by filtering as determined by microscopic fluid analysis following cellular staining. Microscopy of hemocrit containing sample revealed cell types as identified by staining, shape and abundances in accordance with previously reported literature. Amino acid content of hemolymph plasma was compared to whole hemolymph via an electrophoretic assay. Following reaction of hemolymph primary amine components with 3-(4-carboxybenzoyl)quinoline-2-arboxaldehyde electrophoretic separation revealed lower plasma levels compared to whole hemolymph. Amino acid content analysis of hemocrit fraction will also be discussed. Separation of hemocrit from hemolymph demonstrates an ability to further isolate specific Drosophila tissues. The ability to isolate plasma may simplify assay development for hemolymph-mediated chemical signaling whereas hemocyte collection may allow new immunologic experimentation.

467131 AC Electrokinetic Isolation and Detection of Cell Free DNA, RNA and Exosome Biomarkers for Sample to Answer Molecular Diagnostics

Abstract:

Circulating cell free (ccf) DNA, RNA and exosomes are now important biomarkers for liquid biopsy cancer diagnostics, and also hold great promise for early cancer detection. Nevertheless, the isolation of these biomarkers from patient samples requires relatively complex, time consuming and expensive procedures which greatly limits their practical use for most cancer diagnostic applications. New AC dielectrophoretic (DEP) microarray/chip devices now allow 15-20-minute isolation of cancer related ccf-DNA, RNA and exosome biomarkers from 20-50ul of blood, plasma or serum. After isolation of the biomarkers, specific fluorescent dyes can be used first to simultaneously detect the different biomarker levels directly on the chip (in-situ). In a subsequent step, immunofluorescent analysis can be carried out to identify specific protein biomarkers on the exosomes. Finally, the ccf-DNA and RNA (mRNAs and miRNAs release from the exosomes) can be eluted from the DEP chip, and PCR and sequencing analysis carried out to identify the cancer-related point mutations and other polymorphisms, as well as to further verify the tissue origin of the biomarkers. In the case of our Chronic Lymphocytic Leukemia studies, final PCR and DNA sequencing results for CLL Ig heavy-chain variable regions in the IGHV genes and SNPS isolated by DEP were found to be exactly comparable to two much more complex and time consuming â€œgold standardâ€ procedures. In the case of glioblastoma exosomes isolated from plasma, exosome-specific surface and interior proteins CD63 and TSG101 could be detected by immunofluorescence, and mutated EGFRvIII mRNA was detected by RT-PCR. Finally, the exosomal related protein biomarker Glypican-1 could be isolated from pancreatic cancer patient plasma samples by DEP and detected on-chip by immunofluorescence. Thus, DEP represents a powerful new minimally invasive technology for cancer diagnostics that is particularly well suited for the rapid isolation of cell free nucleic acid and exosome biomarkers. The technology is setting the stage for seamless sample to answer liquid biopsy, cancer patient therapy monitoring and ultimately for early disease detection.

468329 In-Depth Single Cell Chemical Analysis Using Nanowell-Based Sample Preparation Combined with Ultrasensitive Capillary Electrophoresis/Mass Spectrometry

Abstract:

The extension of proteomic and metabolomic analyses to the single cell level is an exciting prospect for biological research, as the intercellular heterogeneity that is currently obscured by population-level analyses can be resolved by making chemical measurements one cell at a time. Mass spectrometry (MS)-based single cell analyses require highly sensitive detection as well as essentially lossless sample preparation. We are developing a novel workflow in which cells are robotically dispensed into photolithographically patterned pico- to nanoliter wells, which are covered by an immiscible oil. Requisite sample preparation steps, including cell lysis and proteolytic digestion are performed within the nanowells. The nanowell contents are then analyzed by ultrasensitive capillary electrophoresis-mass spectrometry, or using novel gas-phase mobility based analysis technique known as structures for lossless ion manipulations (SLIM). The current state of the platform and preliminary results using mammalian cell lines will be presented.

469689 Dielectrophoretic Based Molecular Targeting for Blood Chronobiology

Abstract:

The current model of the regulation of the bodyâ€™s circadian rhythm (the 24 hour internal body clock) relies on the rhythmic expressions of â€˜clock genesâ€™ over a 24 hour cycle. These rhythms are of importance and general interest as they are observed in most mammalian cells, *in vivo* and *in vitro*, whilst disruption of these rhythms has been linked to many significant diseases. Work published in 2011 demonstrated, for the first time, rhythmicity in red blood cells (RBCs). Mature RBCs shed their nucleus to maximize their accommodation of haemoglobin and thus are anucleated. Since RBCs contain no nucleus, they are incapable of transcription and thus a cycling gene clock cannot exist in these cells. Very little is known about RBC clock mechanisms, or whether functional consequences exist for erythrocyte biology.

To address this question we have employed the dielectrophoretic-based platform, 3DEP, to characterize isolated RBCs (*in vitro*) as well as whole blood samples (*in vivo*) taken from human and animal subjects. A myriad of experiments has shown robust rhythmicity of the resistive properties (specifically, membrane conductance and cytoplasmic conductivity) of RBCs *in vivo* and *in vitro*. Pharmacological interventions have suggested that this mechanism is at least partially electrophysiologically driven. This work continues in delineating this mechanism through further modifications of the cells *in vitro*.

Additionally, methods have been optimized for the *in vivo* investigation into animal models. This is a significant contribution to the field in not only supporting the *in vitro* work, but since the assay is non-terminal, could greatly reduce animal use in these studies. These assays also serve to give a better understanding into the nature of the rhythms present in whole blood.

469505 Selective Enrichment of Molecular BiomarkersÂ Under Ion Concentration Polarization in Nanochannels Using DC Versus AC Electrokinetics

Abstract:

Selective and rapid enrichment of biomarkers can have transformative impacts within various biosensing paradigms. Affinity methods based on antibody depletion can cause enrichment levels of no more than two to three orders of magnitude. Since typical biomarkers are at million-fold levels lower than interfering proteins or circulating antibodiers, there is a need for complementary enrichment modalities. The current state of the art uses ion concentration polarization in nanochannels to enhance localized fields under DC electrokinetics. Specifically, localized ion depletion at the anodic interface of the microchannel to perm-selective nanochannel is used to enhance electric fields and the resulting biomarker electromigration is balanced against electro-osmosis in the microchannel to cause biomarker enrichment. However, selective enrichment of biomarkers with like-charges, such as prostate specific antigen versus interfering circulating antibodies is not possible and the enrichment levels fall off within physiological media of high conductivity, due to a reduction in ion concentration polarization and electro-osmosis. Herein, we present a comparison of this DC electrokinetic method versus an analogous AC electrokinetic method that also uses a DC offset to locally modulate field profiles by ion concentration polarization. Specifically, by utilizing AC fields at a critical frequency to selectively translate the biomarker of interest by negative dielectrophoresis and a critical offset DC field to create an ion accumulation region proximal to ion depletion along the cathodic interface of the perm-selective region inside a nanoslit, this method seeks to enhance the spatial extent for enabling biomarker enrichment. While enrichment under DC electrokinetics relies solely on ion depletion to enhance fields, the AC electrokinetic mechanism utilizes ion depletion as well as ion accumulation regions to enhance the field and its gradient. Hence, biomarker enrichment under AC electrokinetics continues to be substantial within physiological media, in spite of the lower perm-selectivity. In this manner, by selectively enriching prostate specific antigen versus anti-mouse immunoglobulin antibodies, we demonstrate the elimination of false positives to the immunoassay.

467891 Protein Phosphorylation Cytometry Via Single Cell Isoelectric Focusing

Abstract:

Cellular processes are profoundly affected by chemical modifications of proteins. Perhaps the most important chemical modification is protein phosphorylation, which alters the activity, function, regulation, and degradation of a protein. This broad swath of impact all derives from the simple post-translational addition (or subtraction) of a negatively-charged phosphate group. Given the importance of phosphorylation, it is striking that substantial measurement challenges stymie progress, especially with cellular resolution. Detection specificity (selectivity) is chief among the measurement challenges. When a phospho-specific immunoreagent is available as the basis of an immunoassay, cross-reactivity and off-target binding limit selectivity and multiplexing [2]. Consequently, we advance IEF [1] to measure protein phosphorylation in individual cells.

With single-cell resolution, we profile developmental variation among nominally 'homogeneous' stem cell populations. Through judicious sample preparation, we utilize native and denaturing single-cell IEF to profile both phosphorylated and unphosphorylated protein forms. The rapid (minutes) IEF separation resolves multiple protein species within each cell, down to single-charge unit differences, as is essential for phospho-profiling. After reporting assay and device design, as well as assay characterization, we will overview application of single-cell 'phospho-proteomics' to quantitatively assess and then optimize cell maturation stages within a heterogeneous, stem-cell-derived, terminally-differentiated cell population.

Taken together, this targeted 'phospho-proteomics' assay brings the selectivity of IEF to individual cells, as is central to developmental, disease, and life processes.

References:

1. Tentori AM, Yamauchi KA, Herr AE. *submitted* 2016

2. Bordeaux J, et al. *Biotechniques* 2010.

472691 Preconcentration of Cardiac Troponins in Whole Serum By Isotachophoresis

Abstract:

Microchip isotachophoresis is used to concentrate phosphorylated and unphosphorylated forms of the very low-abundance cardiac troponins (cTnI's) in whole serum while removing many of the high abundance proteins that might interfere with downstream detection. This separation is performed in a "cascade" PMMA microchip that contains a 10x width reduction followed by a 20x depth reduction so that, when ITP is performed with the troponins running in "peak" mode, they may concentrate an additional 200x beyond the ~100x contribution from ITP alone expected for our channel design. We were able to obtain concentration factors as high as 8200 in our experiments and to baseline separate the cTnIs with different phosphorylation states into distinct ITP zones between the leading and terminating electrolytes.

455437 Longitudinal Determination of Vitamin Concentrations in Tears and Blood Serum of Infants and Parents

Abstract:

The viability of utilizing tears for vitamin assessments is investigated. A subset of water-soluble and fat-soluble vitamins were simultaneously determined in tears and blood serum of 52 family pairs; each pair consisting of one infant and one parent. The samples were collected longitudinally during the first year of the child's life. Vitamin concentrations were compared between tear and blood serum, between infant and parent and also against the self-reported nutritional information. Population trends were compiled and quantified using a cross correlation factor. Comparisons of tear vs. blood serum vitamin concentrations revealed no significant difference between tear and serum water-soluble vitamin concentrations, while significantly lower fat-soluble vitamin concentrations were observed in tears compared to serum. Extent of correlation between tear and blood serum vitamin concentrations were then quantified. These results demonstrate that vitamins a) are present and detectable in tears, and b) reflect levels present in blood serum, the current standard biofluid for nutritional screening tests. Infant vs. parent comparisons revealed higher water-soluble vitamin concentrations and lower fat-soluble vitamin A concentrations in infants than parents. Fat-soluble vitamin E concentrations were similar for infants and parents. Dimensions such as bottle-fed and breast-fed were also tracked with infant age and cross-correlated with vitamin concentrations. A higher correlation was found between vitamin concentrations of breast-fed infants and their parents in compared to bottle-fed infants and their parents. Correlations between infant and parent vitamin concentrations decreased longitudinally during the childâ€™s first year and were attributed to the child's diet transitioning from breast-fed to solid food. Vitamin concentrations decreased in both infants and parents during the year of the study in parallel to diet quality assessed via reported dietary intakes. This talk will further discuss concentration results compared against gender, weight, length, head circumference, Apgar numbers, birth complications, excessive/minimal tear production and multivitamin intake. In summary, this study reveals the viability of long-term vitamin monitoring with tears and demonstrates the ability to less invasively screen nutritional health for tracking developmental variations with child growth over time. This work also emphasizes the dependency of infant's nutritional habits to parents.

469983 Improving the Understanding of Early Stage Amyloid Aggregation Using Microchannel Electrophoresis

Abstract:

Amyloids proteins are involved in a variety of diseases including Alzheimerâ€™s Disease (amyloid beta protein) and type 2 diabetes (amylin protein). A number of technical challenges exist for the quantitative analysis of the different sizes of amyloid proteins which are present during the early stages of aggregation, therefore requiring new technologies to be developed for enhanced protein separations. Microchannel electrophoretic techniques have emerged as powerful tools for the quantitative analysis of proteins. Photo-induced crosslinking of unmodified peptides (PICUP) is a powerful tool that can be used to capture the early, transient oligomeric species. While SDS-PAGE has been commonly used to analyze samples that have been subjected to PICUP, here we should the first results of analysis by capillary electrophoresis (CE). In addition, mutations in the protein sequence have been correlated with poorer clinical outcome. Here we should the results of UV-CE analysis of the native protein compared to several mutant proteins.

468402 Refinement of an Internal Standard for Phosphotyrosine Western Blotting

Abstract:

Western blotting using either 1D or 2D sodium dodecyl sulfate polyacrylamide gel electrophoresis (1D or 2D SDS PAGE) in combination with enhanced chemiluminescence detection (ECL) is the most sensitive method known for detecting specific proteins in complex mixtures. Our lab has been using an anti-phospho-tyrosine antibody for 1D and 2D SDS PAGE western blotting to visualize activated tyrosine kinases that might be functioning as cancer drivers in human tumor samples, pTyr-EGFR for example. One difficult problem is quantifying results. ECL light emission varies with time and with western blot conditions. Adding a pTyr-protein standard to every sample should allow normalization of results by determining the density ratio of unknown 1D bands or 2D spots to that of the internal standard. Last year we showed that Alk48 and Alk70 recombinant fragments could be phosphotyrosinylated in vitro with good results, but both pTyr-Alk proteins showed aggregation patterns during 2D gel electrophoresis. Their position was also affected by some samples. In this report, we show that blocking sulfhydral groups by alkylation with iodoacetyamide prevents aggregation patterns of the Alk48 fragment and stabilizes its position. Results are presented for use of the pTyr-Alk48-Iodoacetylated protein as an internal standard.

459344 Comparison of Transport Coal Gasification Process Model Focused on Coal Drying

Abstract:

This work addresses process models for transport coal gasifiers especially in coal drying process. In this work, two process models of the transport gasifier are developed which focus on different coal drying concepts. The first model uses additional gas for drying. The second model imports an integrated drying of feedstock feed to integrated gasification combined cycle (IGCC) plant. In this process, high pressure syngas from the gasifier is heated and is then used to dry feedstock. North Dakota Lignite (NDL) is used as feedstock coal and Peng-Robinson Equation is used for both models as physical property model.

Two simulations are run using the NDL coals and corresponding parameters listed. The ratio of syngas especially H2 and CO determines the latter process. In WTA drying model, ratio of H2 : CO = 1.16 : 1 and itâ€™s suitable for hydroformylation which makes aldehyde (-CHO). In ID case, H2 : CO = 0.49 : 1. Ratio for making SNG is usually 2 : 1 and ratio which is appropriate for methanation is 3 : 1. However, the ratio of ID case is lower for making final product. Therefore, additional process such as Water Gas Shift Reaction (WGSR) process is needed to increase H2 amount.

Two process models of the transport gasifier are developed focused on different coal drying concept. The syngas of WTA model has high temperature and pressure and this energy can be applied in latter process. Integrated Drying concept is suitable for transport gasifier because of high carbon conversion rate and heat stream in gasifier.

459351 Numerical Analysis on Pressure Drop of Distributor in Catalytic Fixed Reactor

Abstract:

In the chemical process, to control a pressure drop is important part in the process. Especially, chimney distributor type largely affects pressure drop and pressure pulsation in the catalytic fixed reactor. So we implement case study according to distributor type. In the case study, a computational fluid dynamics (CFD) model is developed to predict the numerical value of pressure drop in each chimney distributor. In this work, pressure drop is predicted through k-epsilon simulation in the Fluent model. The simulation results can determine whether a venturi inserts or not.

Original distributor, the computational domain was meshed with 382,744 tetrahedric elements and 74,284 mesh nodes. Venturi distributor, it was meshed with 487,148 tetrahedric elements and 94,239 mesh nodes.

The pressure drop ofÂ the original distributor is about 41 pascal, but at the venturi inserted distributor, â–³P is about 2037 pascal. The air-inlet velocity is about 0.377m/s, and oil-inlet velocity is about 0.879m/s. The highest pressure area in chimney is oil-inlet zone. Even though venturi nozzle is inserted to chimney, the mixing effect that has a capability to mix gas with oil through venturi isnâ€™t expected.

The numerical value of pressure drop in two types of chimney distributor is analyzed by the Computational Fluent Dynamics (CFD). For general purpose, original chimney distributor is recommended because the pressure drop is much smaller than venturi inserted distributor. However if mixing effect existed, we have to calculate newly to compare benefit by low pressure drop with benefit by chemical reaction undergone mixing effect.

461458 Meso-Microscopic Analysis of Chemo-Electro-Thermotherapy in Capillary Systems

Abstract:

Electrical field-based technology has been proposed and is used to advance many areas in the medical field. This field of study includes radiology or imaging, neurology, and recently drug delivery, including the field of oncology. Electrochemotherapy is a novel application of this technology that is in need of additional investigation to further conduct effective product design and development within the field. Electrochemotherapy specifically involves the use of a systemic or intratumoral drug injection in conjunction with an electrically generated impulse applied to the specific site of growth. In this effort, a series of fundamental mathematical models will be developed to analyze the role of electrical fields, Joule heating, and interstitial pressure. This main contribution of this work will be to better understand the effects of the order magnitude of the applied electrical field and drug concentration gradient across a capillary bed. This will allow further understanding of drug delivery for the treatment and prognosis of malignant growths. This bio-micro-transport phenomena approach, at the capillary network level, will be followed by computer simulations and data analysis.

461528 Micro-Molecular Scale Modelling of Electro-Chemotherapy

Abstract:

Electrical field-based technology has been proposed and is used to advance many areas in the medical field. This field of study includes radiology or imaging, neurology, and recently drug delivery, including the field of oncology. Electro-chemotherapy is a novel application of this technology that is in need of additional investigation to further conduct effective product design and development within the field. Electro-chemotherapy specifically involves the use of a systemic or intratumoral drug injection in conjunction with an electrically generated impulse applied to the specific site of growth. The electrical impulse is applied in order to increase membrane permeability to allow access of the anticancer drug into the cytosol. After the impulse, the membrane reseals and the anticancer drug exerts its cytotoxicity. The main focus of this contribution is to model with the use of fundamental principles to further investigate the mechanisms by which drug is delivered inside the cellular membrane driven by external forces, such as: the order magnitude of the applied electrical field and drug concentration gradient. The potential role of cellular geometry on the concentration profile of the anticancer drug across the membrane is also investigated. Several cases and scenarios are presented with conclusions for the Electro-chemotherapy treatment.

461359 Effect of Electrical Field and Radius Ratio on the Effective Angular Velocity for Couette and Poiseuille Flows

Abstract:

Different device geometries and flow regimes have been shown to have different effects on the solute separation process of pharmaceuticals and metabolites. Electrical field flow fractionation (EFFF) has been one area of interest to researchers, with a number of parameters being examined. With much of the previous research based on simple geometries, more recent work has expanded the breadth of knowledge to cylindrical devices. The typical approximation of this problem to a rectangular channel has been relaxed to access the role of radius ratios on the velocity field. The main focus of this contribution is to compare the effective angular velocities of Couette and Poiseuille flow regimes in cylindrical EFFF devices. In both situations, due to mathematical complexities, the analytical solution was explored using the spatial area-averaging methodology. This approach will be thoroughly described and several numerical examples will illustrate the most important variations of velocity with the electrical field and radius. The reported findings are expected to assist in product design and development as well as in experimentation.

461720 Molecular Effective Dispersion Under Electrical Field and Channel Curving Effects for Couette and Poiseuille Flows

Abstract:

Among the various geometry schematics for apparatuses utilized in the separation of solutes, including pharmaceuticals and metabolites, cylindrical devices tend to dominate because of their effectiveness and space efficiency. Applications of electrical field flow fractionation (EFFF) has allowed research to be conducted with regard to the optimization of molecular separation by the examination of a few parameters. Micro-channels that exhibit certain degree of curvature influence not only the velocity field but also the degree of solute mixing or separation. To avoid the complexity in mathematical derivations, this phenomenon is relaxed and the effect is mascaraed.

The main focus of this contribution is to broaden the options for apparatus design by investigating the effective diffusivity and optimal time of separation under Poiseuille fluid flow and making the comparison with Couette flow regimes. A theoretical spiraling device was considered in the process. The method of spatial averaging was used in obtaining the analytical expressions of the important parameters. Comparable parameter values were then explored as the correlation was made between the two different flow regimes, ultimately allowing the optimal times of separations to be juxtaposed for conclusion and analysis.

461601 Contact Secondary Nucleation and Self Assembled Monolayers

Abstract:

Understanding and controlling nucleation is a long-standing issue in the field of crystallization and solid state chemistry. In this work we use gold-thiol self-assembled monolayers (SAMs) as hetero-surfaces and induce nucleation by contact force to understand the mechanism of contact-induced heterogeneous nucleation, on different SAMs. A set of contact force nucleation experiments in aqueous glycine solution with SAMs of different functional groups has been performed and compared to a control. Results demonstrate significant reduction in nucleation induction time as well as formation of the beta polymorph of glycine a low supersaturations.

463180 Electrode Topography Effects on Sheared HUVEC Morphology within an Electrical Impedance System

Abstract:

Endothelium dysfunction has been associated with vascular disorders such as atherosclerosis [1], [2] and thrombosis [3], [4]. The structure and function of vascular endothelial cells are regulated by biophysical and biomechanical interactions on both the apical and basal sides. On the apical surface, hemodynamic cues, such as shear stress, control endothelial function and hemostasis [5]. While on the basement membrane, endothelial cell behavior is governed by the topography of the substrate , cellular morphology and the cytoskeleton organization [6], [7]. Electrical impedance spectroscopy provides a real time evaluation of cell migration, cell-substrate and cell-cell interaction [8]. In the study described within, we introduce a microfluidic platform that induces apical and basal-cell interactions and replicates the length scale and structure of a blood vessel, with the ability to electrically characterize endothelial cell behavior using an embedded impedance sensor. Apical cues are simulated by subjecting Human umbilical vein endothelial cells (HUVECs) to a constant shear stress of 15 dyne/cm^2 and a stepped shear condition of 20-30-50 dyne/cm^2, while, basement membrane cues are generated by seeding HUVECs into microchannels containing topographical features pertaining to the electrodes. Initial results indicate that both apical and basal cues direct. cell morphology adaptation. While cells align and stretch in the direction of flow on areas of smooth surface, they tend crowd together and arrange themselves along topographical features. Understanding how endothelial cell morphology is directed by the interaction between hydrodynamic shear stress and topographical features of electrodes within an impedance system is important in addressing these effects in impedance measurements and the need for improvements in *in vitro* impedance platforms to evolve into more physiologically relevant models. Overall, these studies and technology improvements can be crucial in further comprehending normal and aberrant endothelial cell function, which can lead to new discoveries on how to treat and prevent vascular diseases.

References:

[1] P. O. Bonetti, L. O. Lerman, and A. Lerman, "Endothelial dysfunction: A marker of atherosclerotic risk," *Arterioscler. Thromb. Vasc. Biol.*, vol. 23, pp. 168–175, 2003.

[2] P. M. Vanhoutte, "Endothelial dysfunction and atherosclerosis," *Eur. Heart J.*, vol. 18, pp. 19–29, 1997.

[3] K. K. Wu and P. Thiagarajan, "Role of endothelium in thrombosis and hemostasis.," *Annu. Rev. Med.*, vol. 47, pp. 315–331, 1996.

[4] J. E. Deanfield, J. P. Halcox, and T. J. Rabelink, "Endothelial function and dysfunction: testing and clinical relevance.," *Circulation*, vol. 115, no. 10, pp. 1285–95, Mar. 2007.

[5] J. T. Morgan, J. a Wood, N. M. Shah, M. L. Hughbanks, P. Russell, A. I. Barakat, and C. J. Murphy, "Integration of basal topographic cues and apical shear stress in vascular endothelial cells.," *Biomaterials*, vol. 33, no. 16, pp. 4126–35, Jun. 2012.

[6] D. E. J. Anderson and M. T. Hinds, "Endothelial cell micropatterning: Methods, effects, and applications," *Ann. Biomed. Eng.*, vol. 39, no. 9, pp. 2329–2345, 2011.

[7] P. Lu, V. M. Weaver, and Z. Werb, "The extracellular matrix: a dynamic niche in cancer progression.," *J. Cell Biol.*, vol. 196, no. 4, pp. 395–406, Feb. 2012.

[8] I. Giaever and C. R. Keese, "Micromotion of mammalian cells measured electrically.," *Proc. Natl. Acad. Sci. U. S. A.*, vol. 88, no. 17, pp. 7896–900, Sep. 1991.

463208 A Framework for Undergraduate Research on Chemo-Electro-Thermotherapy

Abstract:

Research is known to be much more challenging at the undergraduate than graduate level. Lack of readiness and/or knowledge are good examples (sometimes excuses) as to why it is not very appealing to conduct research in collaboration with undergraduate students. However, undergraduate research is greatly rewarding, specifically to students planning on continuing their education after graduation. Early introduction to research, at the undergraduate level, must compass a review of the state of the art of the phenomena under study with a holistic view of practical applications and fundamentals involved in describing it.

The primary objective of this contribution is to report the main finding after applying the state of the art approach to the subject of Chemo-Electro-Thermotherapy. The report includes multiple categories: Cancer and the major effects of Angiogenesis, Tumor Growth and Electrical Fields, DNA and Enzymatic Response to Cancer, and Nanotherapeutics. Subcategories have been thoroughly examined with special interest in mapping fundamental principle equations. Furthermore, a series of fundamental mathematical models will be discussed with respect to their potential contribution to treatment and prognosis of malignant growths. Lastly, this contribution intends to show a framework to identify research opportunities in the area of electrical field, Joule heating, interstitial pressure and carboplatin, to name a few, from bio-micro-transport phenomena approach.

463217 Dynamic Modeling of Fluid Flow Fractionation Under Couette and Poiseuille Flows

Abstract:

Parameters such as effective velocity and effective diffusivity capture the efficiency of species separation. The study of how different physical conditions affect these parameters is relevant. The effects of orthogonal fields and radius ratios on the effective parameters were analyzed and developed in a previous effort (Dueck, Oyanader & Oyanader, 2015). These equations can be implemented in the development of a concentration profile of a solute within a cylindrical and/or rectangular cell system. It is traditional that the dynamic concentration profile in a typical FFF cell was modeled using the Dirac delta function in the representation of the solute injection, giving it an initial concentration value that infinite. This function is often used to simplify pulse like mathematical equations, but did not accurately embody the actual behavior of the system.

The equation for the dynamic behavior of the concentration of species is modelled omitting the use of the Dirac delta function but including a forcing step expression. The new equation represents more realistically the phenomena and the results of a pulse injection of solute into the system. Six main variables affecting the separation process were investigated. To close, the results and conclusions derived from this study will be highlighted to illustrate the modeling technique as well as the system behavior.

469517 Nano-Templated Agarose-Gels for Tissue Scaffolds: Preliminary Synthesis and Characterization of Internal Structure and Transport Behavior

Abstract:

Hydrogel scaffolds have tremendous applications in the field of healthcare engineering including such areas as tissue engineering and controlled transport of drug and cell therapies. Their extremely high hydrophilicity and permeability to solutes through their porous structure makes these materials excellent choices for development. Tissue engineering, in particular, is an interdisciplinary and multidisciplinary field of regenerative medicine that aims at a development of biological substitutes to restore, maintain, and improve impaired tissue functioning. Agarose gel-based scaffolds are excellent potential candidates for this purpose. This biomaterial can be fabricated, modified, or incorporated with other materials to achieve and enhance desired transport and mechanical properties. In this research project, we are focused on the manufacturing and characterization of transport performance of nano-templated gels based on agarose. More specifically, we are interested in determining the influence of modifying the gel microstructure on the rate of solute (nutrient) transport by placing and subsequently removing templates before completing transport studies. The core techniques to change the internal structure of these hydrogels are: (1) changing the concentration of agarose and (2) incorporating suitable nano- or micro- sized templating agents during polymerization. Towards this end, microstructure modifiers (e.g., nano-particles) will be incorporated and transport is examined in the absence or presence of applied electrical fields induced by running gel electrophoresis for modified and standard agarose gels. Ultimately, gel-electrophoresis is performed for studying electro kinetic rates of nutrient transport. Thermal techniques (e.g., thermal gravimetric analysis and differential scanning calorimetry) and electron microscopy (e.g., transmission electron microscopy and scanning electron microscopy) are the instrumental techniques being examined for characterization of internal structures such as porosity, pore shape, and size distribution. The goal is to produce agarose-based hydrogels with optimal properties to achieve desired transport and mechanical properties.

471741 Following Lineage Commitment of Pre-implantation Embryos Through Single-Embryo Western Blotting

Abstract:

Before implantation into the uterine wall, embryos undergo multiple cell divisions and transitions from zygote (1-cell) to 2-cell, 4-cell, 8-cell, morula and finally blastocyst (3). The first cell lineage specification occurs in these pre-implantation stages, with blastomeres dividing to form the inner cell mass (ICM) and the trophectoderm (TE) at the blastocyst stage. Debates remain whether this first asymmetry is due to different molecular phenotypes in early blastomeres, or whether cells are assigned to the ICM or TE depending on their position and polarity at the 32-cell stage (2). To answer these questions, research on pre-implantation development has typically focused on gene expression levels. However, the fluctuations in RNA caused by transcriptional bursts does not often correlate well with protein levels, given the different half-lives of RNA and protein (2). The need to perform protein measurements on single embryos and single blastomeres remains.

To enable detection of proteins from embryos, we have advanced the single cell western blot (scWestern), recently developed in the Herr Lab, to analysis of whole embryos. The scWestern is a thin polyacrylamide gel layer, grafted onto a standard microscope slide, and stippled with thousands of microwells (1). Cells are settled into microwells, lysed in-situ, and solubilized proteins are electrophoresed across the polyacrylamide layer. Proteins are then covalently bound to the polyacrylamide, by activation of photoactive molecules (benzophenone moieties) incorporated into the polyacrylamide matrix. Fluorescently-labeled antibodies are then diffused into the polyacrylamide for detection of protein targets. By increasing the microwell dimensions and polyacrylamide thickness to accommodate embryos (200 um in diameter), modulating the pore size to avoid clogging at the gel interface, and introducing lysis through a lid system, we have achieved simultaneous analysis of all pre-implantation stages, from zygote to blastocyst. We have detected

bands of loading controls and are now focusing on markers associated with totipotency and early differentiation. Furthermore, to help answer whether single blastomeres show signs of lineage commitment before the 32-cell stage, embryos will be dissociated into individual blastomeres and analyzed individually for markers of TE and ICM. Protein analysis of embryos and single blastomeres will result in detailed temporal tracking of the emergence of cell asymmetry, helping to confirm or challenge current evidence supported by transcriptomic data.

References:

1. Hughes et al. "Single cell western blotting". Nature Methods. 2014. 11(7): 749-55.

2. Nimmo et al. "Primed and ready: understanding lineage commitment through single cell analysis". Trends in Cell Biology. 2015. 25: 8; 459-467.

3. Welling et al. "Symmetry breaking in the early mammalian embryo: the case for quantitative single-cell imaging analysis". Mol. Hum. Reprod. (2016) 22 (3): 172-181.

478572 Numerical Model of Streaming DEP for Stem Cell Sorting

Abstract:

Neural stem cells are of special interest due to their potential in neurogenesis to treat spinal cord injuries, in promoting brain repair and other nervous disorders.[1,2] Current common methods used to quantitatively characterize stem cells include fluorescence activated cell sorting, (FACS) [3], magnetic bead-coupled cell separation[4] and micropipette aspiration [5]. These are limited due to the lack of antigens and labels that are specific enough to stem cells of interest. Dielectrophoresis is a label free separation technique that has been recently demonstrated for the enrichment of neural stem/progenitor cells using the membrane capacitance of the cell as the distinguishing factor[6]. Here we use numerical simulation to investigate the use of streaming DEP for the continuous sorting of neural stem/progenitor cells. The aim is at understanding how select device and experimental variables affect the throughput and efficiency while continuously sorting SC27 stem cells, neurogenic progenitor, from SC23 cells, an astrogenic progenitor. The sorting is studied by characterizing the width of the stream obtained for these cells with a purity ≥98% for continuous separation. The variables studied here are electrode cross section shape (circle, lens and diamonds), height (10, 50 and 100% of the channel height), the flow rate (200-2000 µl/min) and the cell concentration (10^3 to 10^5 cells/ml). Based on the results of the simulation, a device is proposed to retrieve the streams.

1. Martino, G.; Martino, G.; Pluchino, S.; Pluchino, S. The therapeutic potential of neural stem cells. ***Nat. Rev. Neurosci.*** **2006**, *7*, 395–406.

2. Okada, S.; Ishii, K.; Yamane, J.; Iwanami, A.; Ikegami, T.; Katoh, H.; Iwamoto, Y.; Nakamura, M.; Miyoshi, H.; Okano, H. J.; Contag, C. H.; Toyama, Y.; Okano, H. In vivo imaging of engrafted neural stem cells: its application in evaluating the optimal timing of transplantation for spinal cord injury. ***FASEB J.*** **2005**, ***19***, 1839–1841.

3. Uchida, N.; Buck, D. W.; He, D.; Reitsma, M. J.; Masek, M.; Phan, T. V; Tsukamoto, A. S.; Gage, F. H.; Weissman, I. L. Direct isolation of human central nervous system stem cells. ***Pnas*** **2000**, ***97***, 14720–5.

4. Spangrude, G. J.; Heimfeld, S.; Weissman, I. L. Purification and characterization of mouse hematopoietic stem cells. ***Science (80-.).*** **1988**, ***241***, 58–62.

5. Hochmuth, R. M. Micropipette aspiration of living cells. ***J. Biomech.*** **2000**, ***33***, 15–22.

6. Flanagan, L. A.; Lu, J.; Wang, L.; Marchenko, S. A.; Jeon, N. L.; Lee, A. P.; Monuki, E. S. Unique Dielectric Properties Distinguish Stem Cells and Their Differentiated Progeny. ***Stem Cells*** **2008**, ***26***, 656–665.

479013 A Microfluidic Device for Low-Input Methylomic Analysis Based on Reduced Representative Bisulfite Sequencing

Abstract:

Methylomic studies require substantial amounts of DNA samples and this restriction hinders applications involving scarce animal or patient samples with direct biomedical relevance. Here we report a microfluidics-based reduced representative bisulfite sequencing (RRBS) protocol that permits methylomic profiling with sub-1 ng starting DNA. We used diffusion-based reagent swapping for the multi-step treatment and purification of DNA on the microfluidic platform. Our device and protocol effectively minimized the loss of function for DNA to $< 70\%$ while achieving high bisulfite conversion within a 2 h period (in comparison to up to 96% DNA function loss in conventional bisulfite conversion protocol due to DNA hydrolysis under acidic condition). Using this technology, we studied DNA methylation in neurons and glia isolated from mouse cerebellum, revealing cell-type specific methylomic patterns and their influence on gene activities.

479306 On-Chip Separation of Triangular and Cylindrical DNA Origami Nanostructures Using Slanted Nanofilter Array

Abstract:

DNA origami (DO) has enabled design and fabrication of DNA nanostructures of a variety of shape and functions. Based on such features, many novel DO molecules have been devised as a molecular probe to detect important target molecules such as disease related protein or miRNA, which is designed to undergo certain configuration/shape change in the presence of the target molecules [1]. Such reconfiguration of DO has been typically detected with other instruments such as AFM and single molecular FRET. However, for practical applications such as point-of-care (POC) and detection of low concentration target molecule, more rapid and less costly methods with improved detection limit are demanded. At the same time, many on-chip micro/nanofluidic separation methods have been developed recently, enabling rapid and continuous separation of differently shape molecules with high resolution. By utilizing such on-chip separation devices, it is expected that DO of different shape/configuration is rapidly separated and reconfiguration efficiency is easily analyzed.

In this work, as a first step toward the establishment of such on-chip DO based molecular detection system, the difference in electrophoretic migration is characterized for two differently shaped DOs, triangular DO and cylindrical DO, inside a recently reported nanofluidic sieving device called Slanted Nanofilter Array. Specifically, the mean and STD of deflection angle for each DO were measured, varying device parameters such as electric field strength and nanochannel depth. In Slanted Nanofilter Array, the deflection angle of each molecule is uniquely determined by its size and shape based on Ogston sieving mechanism, and the separation resoultion and detection limit is greatly increased by the integrated preconcentration region in the device [2]. Subsequently, a rigid body Brownian dynamics simulation was performed to compare with the experimental result, and it was shown that the experimental result could be well reproduced in the simulation both quantitatively and qualitatively. Furthermore, in order to increase the accuracy of the simulation, electrohydrodynamic parameters such as electrophoretic mobility and diffusion coefficient were measured with conventional CZE for each DNA origami nanostructure. Accordingly, the experimental results gained in this work are expected to be vital information for the design of an on-chip DO based molecular detection system.

479867 Selective Catalytic Reduction of Nitric Oxide By Ammonia over V2O5/ TiO2 in a Hollow Cylindrical Catalyst Under Enhancing Effect of Electrohydrodynamics: A Kinetic Modeling Study

Abstract:

Emissions of nitrogen oxides (NOx) in fuel combustion from stationary and mobile sources contribute to the greenhouse effects and cause a variety of environmentally harmful results. Meanwhile, the selective catalytic reduction (SCR) of NOx using ammonia over a vanadia-based catalyst in heterogeneous reactors is still a well-proven technique for NOx abatement. However, to meet forthcoming restrictive legislation, greater De-NOx performance can be achieved by eliminating operational setbacks such as diffusion resistance, poor catalyst activity at low temperatures, etc. In this study, an electrohydrodynamic (EHD)-SCR model was developed to evaluate the enhancing effect of the EHD technique on NH3-SCR of NO through a hollow cylindrical catalyst. Computational investigations were performed based on the proposed model in different operational conditions to examine the effect of various operating parameters on SCR enhancement. Simulation results showed that catalyst utilization was intensified significantly by EHD application and generation of additional flow known as EHD-induced secondary flow through the catalyst porous layer, which undermined the proposed drawbacks of the catalyst medium and provided higher catalyst effectiveness with greater NO conversion. Results also indicated that the maximum enhancement of almost 4.2-fold could be obtained for NO conversion with electric potentials and operating temperatures of 150â^'270 V and 150â^'165 Â°C, respectively.

479870 Synthesis, Purification, and Capillary Electrophoretic Separation of Fluorescent Carbon Nanodots

Abstract:

Fluorescent carbon nanodots (C-dots) have drawn considerable attention in recent years. C-dots are carbon nanoparticles with typical sizes of < 10 nm and have been reported to have excellent characteristics (e.g., low cost, colloidal stability, upconversion photoluminescence, reduced cytotoxicity, very good biocompatibility) that make them promising candidates for a wide range of applications, such as bioimaging, biolabeling, fluorescent ink, photocatalysis, and optoelectronic devices [1]. After the discovery of C-dots [2], much research efforts have focused on developing better synthetic routes to obtain C-dots with relatively high quantum yield (QY). Certainly, numerous synthetic methods have been developed, and fluorescent products with relatively high QY have been reported. However, the purification of the fluorescent products is often neglected and it is difficult to discern if indeed the reported photoluminescence is the result of C-dos or other potential molecular species that can be byproducts of the synthetic route. Therefore, one must ask: are all the photoluminescent species in the synthetic product C-dots? Are the commonly used purification methods good enough to completely isolate the C-dots from potential small fluorescent molecules?

In addressing these questions, we are investigating how the purification process affects the purity of fluorescent C-dots. We have used one of the most commonly reported approaches to synthesize C-dots, namely the hydrothermal treatment of simple molecular precursors. We used citric acid and 1,2-ethylenediamine as the molecular precursors in the production of C-dots. The as-synthesized fluorescent product was purified by the most commonly used purification method reported in the literature; that is the dialysis against pure water [3,4]. The fluorescent products inside the dialysis membrane (retentate) and that outside the membrane (permeate) were collected at different time intervals during the dialysis process that proceeded for several days. The QY of each sample was determined. In addition, the samples were analyzed via capillary electrophoretic (CE) and laser induced fluorescence (LIF). The CE separations

showed variation in complexity of retentate and permeate during the dialysis process and the QY was also different. The details of our experiments and findings will be the focus of this presentation.

References

1. Baker, S. N.; Baker, G. A. *Angew. Chem., Int. Ed.* **2010**, *49*, 6726-6744.
2. Xu, X. Y.; Ray, R.; Gu, Y. L.; Ploehn, H. J.; Gearheart, L.; Raker, K.; Scrivens, W. A.; *J. Am. Chem. Soc.* **2004**, *126*, 12736-12737.
3. Zhu, S.; Meng, Q.; Wang, L.; Zhang, J.; Song, Y.; Jin, H.; Zhang, K.; Sun, H.; Wang, H.; Yang, B. *Angew. Chem., Int. Ed.* **2013**, 52, 3953-3957
4. Zhai, X.; Zhang, P.; Liu, C.; Bai, T.; Li, W.; Dai, L.; Liu, W. *Chem. Commun.* **2012**, 48, 7955-7957.

479872 Clarification of Breast-Cancer Infected Peripheral Mononuclear Blood Cells in a Semi-Circular Insulator-Based Microfluidic Channel

Abstract:

Currently, early stage breast cancer detection is a major challenge faced by developed and developing countries. This work presents a simulation approach to characterize and separate breast cancer infected peripheral blood mononuclear cells (iPBMCs) from the healthy mononuclear cells (PBMCs) using peripheral blood stream via electrodeless dielectrophoresis (eDEP) on a microfluidic platform. COMSOL Mutiphysics v5.2a was used to simulate the device geometry consisting of a microchannel incorporating semicircular obstacles and bifurcating into two outlet channels with different applied voltages and different inlet concentrations in the range of 1 iPBMC per 10 - 108 healthy PBMCs. Mathematical models were solved for current conservation along with fluid flow, convection, and diffusion to predict the concentration distribution of the cells. Flux streamlines were used to represent the path followed by iPBMCs and PBMCs. Results indicate an optimal separation at 30 V and a recovery rate of ~ 89% irrespective of the concentration of the iPBMCs fed into the device. A micro well was used to obtain the electrophysiological properties of iPBMCs and PBMCs and these were used in optimizing the device.

479873 Can a Neutral Particle Translate in an Electric Field?

Abstract:

We introduce the self energy of mobile ions to the standard continuum model of electrokinetic processes. The self energy captures ion solvation, ion-ion correlation, and spatial permittivity variation effects in inhomogeneous dielectric systems. Differences in anion and cation self energies can engender local charge separation and hydrodynamic flow not observed in the mean-field theory of electro-osmosis.

We illustrate such behavior for a semi-infinite plate geometry. The Helmholtz-Smoluchowski formula for electro-osmotic slip is recovered; importantly, the mean zeta potential is self-consistent with an equilibrium charge distribution that also depends on ion self energies. If the surface is neutral, then classical theory anticipates the potential and fluid velocity are everywhere zero. We predict, however, that a valence asymmetric salt solution (e.g. $CaCl_2$) adjacent to a neutral surface will induce a *non-zero* zeta potential and, therefore, a *finite* slip velocity. This unintuitive result suggests a finite electrophoretic mobility for a uniformly neutral, non-polarizable particle in an asymmetric salt solution. Our analysis challenges the present understanding of the phoretic particle charge-mobility relationship, namely coincidence of zero charge and zero mobility. So yes, under certain conditions, a neutral particle can translate in an electric field!

478738 Assessment of Joule Heating Effects in Optimized Insulator Based Dielectrophoresis Devices

Abstract:

Insulator-based dielectrophoresis (iDEP) has been widely used for particle and cell manipulation. It relies in the inclusion of insulator structures between two remote electrodes to create non-uniform electric fields and induce polarization effects into the particles to produce a net movement. If this force overcomes all other forces in the system, trapping can be achieved. Usually, this requires high electric potentials, especially if the particles are small in size, which increases temperature due to Joule heating, having setbacks such as a possible damage to the sample, bubble formation, undesired electro-thermal effects, etc. To improve performance of iDEP microdevices, optimization of the insulatorâ€™s shape and size has been performed which increased the dielectrophoretic effect by increasing the gradient of the electric field square. In this work, 3D mathematical simulations were developed in order to validate the optimized microdevices when Joule heating effects are considered. These results can open a new perspective of a balance between increasing the dielectrophoretic force to improve manipulation of particles and cells, without increasing the system temperature to undesired values.

478766 *The Impact of Different Waveforms on Particle Trapping Efficiency When Using 3D Carbon-Electrode Dielectrophoresis*

Abstract:

Here we study the use of different waveforms when trapping latex particles on 3D carbon electrodes contained in a microfluidic channel using positive dielectrophoresis (DEP). DEP refers to motion induced on a targeted particle by a non-uniform electric field. The magnitude of such motion depends mainly on the particle size and the magnitude of the polarizing signal. The direction of the force, either towards as in positiveDEP or away from the field gradient in the case of negativeDEP, depends on the difference of electrical polarizability between the particle and the suspending media. Here we focus on assessing the potential advantages of optimizing the waveform of the polarizing signal to improve efficiency when extracting particles from a flow. Sinusoidal, square, and ramp signals of a given frequency and amplitude were characterized and compared based on their RMS voltage value (see table 1). The RMS value of the square signal is equal to its amplitude *a*, while the RMS values of the sinusoidal and ramp signal are only and of the signal amplitude respectively. Hence, the use of a square waveform is expected to induce the strongest electric field gradient of the three waveforms, and thus a higher degree of trapping. The goal of this work is to quantify the trapping latex particles when using different waveforms of varying frequencies.

Fluorescent latex particles, 1µm in diameter, were suspended in Distilled Water (DI) to obtain a concentration of 10^7 particles/ml. The experimental protocol is similar to that previously presented by this group [1-4]. Briefly, the particles were flowed through an array of 3D carbon electrodes polarized by a specific waveform to trap targeted particles. Once trapped, the particles were washed with DI water. The field was then turned off to release the particles. The fluorescence intensity at the end of the electrode array was monitored throughout the experiment. Trapping efficiency was evaluated by comparing the difference in fluorescence intensity between the frame immediately before turning the field off and the area under the intensity curve obtained after turning the electric field off. Hence, a high area value means strong trapping. ImageJ software was used to obtain the levels of brightness in each frame of the recorded video. The data was plotted and the sums of the values under the curve were compared. The frequencies of interest ranged from 5 kHz to 50 kHz at an amplitude of 15 Vpp.

Initial results are shown in Figure 1. Using the square signal, the greatest amount of particles was released when the field was turned off compared to the sinusoidal and ramp signals at all frequencies above 5 kHz. This behavior suggests that trapping is most effective when using the square signal. The amount of particles trapped is proportional to the RMS value of each signal. Because the RMS value of the square signal is the greatest among the sinusoidal and ramp signals, it trapped the most particles on the electrodes. Likewise, the ramp signal, with the lowest RMS value, was the least effective at trapping particles on the carbon electrodes. Electrothermal effects begin to heat the media at low frequencies and cause the particles to waver between electrodes; thus, results below 5 kHz were inconclusive.

From these experiments the use of a square signal appears to be the most effective in trapping latex particles when compared with the sinusoidal and ramp signals. The results produced coincide with the values of the RMS value of each electrical signal.

Ongoing work is on characterizing trapping based on the energy delivered by each of the signals and the dynamics of particle accumulation in positiveDEP areas when using each of the waveforms. This will permit a better understanding on the phenomena induced when using different signals.

REFERENCES:

1. M. Islam, R. Natu, M. F. Larraga-Martinez and R. Martinez-Duarte "Enrichment of diluted cell population from large sample volumes using 3D Carbon-electrode Dielectrophoresis" *Biomicrofluidics*, 10, 033107 (2016).
2. M. Elitas, R. Martinez-Duarte, N. Dhar, J. McKinney and P. Renaud, "Dielectrophoresis-based purification of antibiotic-treated bacterial subpopulations". *Lab on a Chip*, 14, (11) 1850-1857 (2014).
3. M.C. Jaramillo, R. Martinez-Duarte, M. Hüttener, P. Renaud, E. Torrents and A. Juarez, "Increasing PCR sensitivity by removal of Polymerase inhibitors in natural samples using Dielectrophoresis". *Biosensors and Bioelectronics*, 43, 297-303 (2013).
4. R. Martinez-Duarte, F. Camacho-Alanis, P. Renaud and A. Ros, "Dielectrophoresis of lambda-DNA using 3D carbon electrodes" *Electrophoresis*, 34, 1113-1122 (2013).

479874 Characterizing Human Stem Cell Function with Dielectrophoresis and Flow Cytometry

Abstract:

Human embryonic stem cells (hESCs) provide great opportunities in stem cell therapeutics because they can differentiate into the three germ layers: ectoderm, mesoderm, and endoderm. More specifically, hESCs can be directed toward human neural stem/progenitor cell (hNSPC) differentiation and used to treat neurological diseases and injuries. Sufficiently characterizing hNSPC's functional behavior before using them in transplant therapy is essential to the development of reliable therapeutic treatment options. Having measures that will accurately reflect cell phenotype after transplantation is critical. Currently, the cell characterization process is challenging due to a lack of biomarkers that provide an adequate picture of the specific functions of hNSPCs. Therefore, we've implemented dielectrophoresis (DEP), a label-free characterization technique that uses nonuniform electric fields, to determine cell dielectric properties such as membrane capacitance (C_{mem}). Additionally, cells were analyzed by flow cytometry to characterize hNSPC surface protein expression to determine any correlations between characteristics measured by DEP and cell surface markers.

In this work three sets of hESC-derived hNSPCs (Shef4-1, Shef4-2, Shef4-3) were derived from passage 6 EZ spheres and established as monolayers before analysis. The Shef4-1, Shef4-2, and Shef4-3 nomenclature indicates that these cells were derived from the same set of EZ spheres but separately differentiated to form hNSPCs, as would be done to generate sufficient numbers of stem cells for therapeutic purposes. Once established as monolayer hNSPC cultures, their cell size, DEP spectra, C_{mem}, and cell surface protein expression were quantified at a variety of passage numbers to determine lot-to-lot variability or variability over passaging. Results show that cell diameter varies across lots, so Shef4-1 > Shef4-2 > Shef4-3. Similarly, there is variability across lots in the DEP spectra, which is corroborated with Cmem. At passage 7 between 0-100kHz the DEP spectra is shifted right for Shef4-2 (C_{mem} = 8.6mF/m^2) as compared to Shef4-1 and Shef4-3 (C_{mem} = 16.5mF/m^2 and 16.4mF/m^2, respectively). Over passages, Shef4-2 C_{mem} decreases with passage number, Shef4-3 C_{mem} increases with passage number and Shef4-1 does not have a clear C_{mem} trend. Cell surface markers were assessed by flow cytometry resulting in alpha6 and alphaV integrin expression decreasing over passage for Shef4-1, increasing for Shef4-2, and decreasing for Shef4-3. The changes in alpha6 and alphaV integrin expression over passage correlate well with changes in C_{mem} over passage. Generally, these findings correspond well with DEP theory such that high alpha6 and alphaV expression indicates a complex cell surface, which translates to lower C_{mem} (Shef4-2 C_{mem} decreases and integrin expression increases over passage) with the opposite also being true (Shef4-3 C_{mem} increases and integrin expression decreases over passage).

Here we've shown that significant lot-to-lot variability exists among hESC-derived hNSPCs, and DEP plus flow cytometry provide a good quantitative assessment of cell phenotype variability. Additionally, effectively discerning the functional behavior of stem cells before their use in transplants is essential to advance stem cell therapeutics; alpha6 and alphaV integrin are important for cell migration and differentiation. Linking integrin protein expression with DEP measurements provides additional insights to the biological meaning of C_{mem} to expand beyond the core definition, which is the ability to store charge. Future work will compare transplantation efficiencies of different hESC-derived hNSPCs.

470685 Origin and Nature of Charge Carriers in Nonpolar Solvents

Abstract:

Using either electrochemical impedance spectroscopy [1] or a commercial meter, we measured conductivity of various commercial surfactants in dodecane for surfactant concentrations between 1 and 100 mM. For Span85, AOT, Span80, OLOA and Span20, the conductivity was found to be proportional to surfactant concentration and the charge carriers are micelles of the surfactant. The proportionality constant (conductivity divided by the molar concentration of surfactant) increases exponentially with the size of the micelle. This is surprising because larger micelles should be less mobile than smaller ones. The explanation is that the fraction of micelles that bear a charge (which is always quite small compared to unity) increases exponentially with the size of the micelle. Bigger micelles mean that opposite charges in ion-pairs are further apart, leading to a lower attractive energy compared to kT and a greater degree of dissociation of the ion pairs.

These observations suggest that larger, wormlike micelles should be highly potent charging agents. To test this, we have synthesized a series of phospholipid-derived surfactants known to form wormlike micelles in nonpolar solvents. In this presentation we will present their concentration-dependent conductivity enhancement in nonpolars and compare it to the charge-fluctuation theory developed by Eicke and coworkers [2]. We have also synthesized a library of AOT-like surfactants with varying head and tail group chemistry and counterions. These materials have very different chemical composition but form micelles of similar size. Conductivity and impedance spectroscopy measurements on each reveal the relative importance of micelle size and chemical composition on their effectiveness as charging agents, and suggest mechanisms for the origin of charge formation in doped nonpolar liquids. These will also be discussed along with a consideration of the role of adventitious water in these systems.

[1] B.A. Yezer, A.S. Khair, P.J. Sides and D.C. Prieve, J. Colloid Interface Sci. (2014), dx.doi.org/10.1016/j.jcis.2014.08.052.

[2] Eicke H-F, Borkovec M, Das-Gupta B. Conductivity of Water-In-Oil Microemulsions: A Quantitative Charge Fluctuation Model. J Phys Chem. 1989;93:314-7.

455914 Isomotive Dielectrophoresis (isoDEP): Characterization through Particle Velocemitry

Abstract:

Dielectrophoresis (DEP) is the translation of a polarizable particle when subjected to a non-uniform electric field. Most DEP devices focus on trapping or sorting particles and overlook its analytical capabilities. This work describes a parallel cell analytical system utilizing an isomotive dielectrophoresis (isoDEP) environment. With this approach detailed dielectric measurements of individual particles are possible.

The DEP force of a homogeneous spherical particle is $F_{DEP}=2\pi\varepsilon_m a^3 Re[f_{CM}](gradient(E^2))$ where ε_m is the permittivity of the fluid medium, *a* is the particle radius, f_{CM} is the Clausius-Mossotti factor, *E* is the electric field. The f_{CM} governs the magnitude and direction of DEP-induced motion and is a function of particle composition. The imaginary component of f_{CM} can be measured with electrorotation (ROT) [1,2] and the resultant rotational rate is a function of AC frequency as well as the dielectric properties of the media and cell. ROT's most significant disadvantage is its limited throughput [1]. Further, a common characteristic of existing DEP electrode geometries is that the DEP force is significantly non-uniform in the sample region due to the gradient of the field-squared ($gradient(E^2)$), making parallel analysis of cells non-trivial.

The unique characteristic of isoDEP is that the experimental region has a *constant* value of $gradient(E^2)$ resulting in a uniform DEP force inducing a constant (isomotive) particle translational velocity. Inspired by initial analysis by Herbert Pohl [3], we have developed a modified electrode geometry for isoDEP. Fabrication of extruded electrodes is straightforward via microfabrication methods (DRIE of conductive wafers) or sub-millimeter machining. A sample is injected and flow is halted before field activation. Digital images will extract particle size and, due a constant $gradient(E^2)$, the only unknown for each particle is f_{CM}. The field is applied and f_{CM} is extracted through particle tracking. The particle's velocity

will change as the AC frequency is swept over a specified range to obtain a comprehensive f_{CM} spectrum. Through simultaneous particle tracking such spectra are obtained for every particle in the imaging area, enabling parallel analysis of particles. In the future, isoDEP will be used to extract the dielectric properties of each cell (ex: membrane capacitance) – these properties directly correlate to the cell physiology.

References

[1] S.I. Han, Y.D. Joo, and K.H. Han, "An electrorotation technique for measuring the dielectric properties of cells with simultaneous use of negative quadrupolar dielectrophoresis and electrorotation", Analyst, 138, 1529, (2013).

[2] M. Cristofanilli,, G. De Gasperis, L. Zhang, M.-C. Hung, P.R.C. Gascoyne, and G.N. Hortobagyi, "Automated electrorotation to reveal dielectric variations related to HER-2/neu overexpression in MCF-7 sublines", Clinical Cancer Research, 8, 615, (2002).

[3] H.A. Pohl, *Dielectrophoresis: The Behavior of Neutral Matter in Nonuniform Electric Fields*, Cambridge; New York: Cambridge University Press (1978).

457942 A Novel Electrokinetic-Electromechanical Microfluidic Platform Using Conductive Carbon Black Membranes

Abstract:

Conductive materials have been widely utilized and integrated into microfluidic devices for on-chip electrokinetic applications. While the high electrical conductivity of metal films such as gold, chrome and indium tin oxide (ITO) makes them suitable for delivering electric field to fluid contained within these microfluidic chips, these structures are expensive to fabricate and are typically confined to the microfluidic channel surface, which limits their ability to electrokinetically manipulate the flow volume. On the contrary, soft materials like agarose gel and hydrogels are cheaper and easier to fabricate into 3D structures. Their poor mechanical strength, however, often leads to unreliable devices with unstable electrokinetic performance. In this talk we present a new 3D microfluidic platform based on conductive carbon black nano-composite PDMS. By locally patterning conductive PDMS directly into the sidewalls of a microfluidic flow channel, we demonstrate the ability of this platform to perform a wide variety of on-chip electrokinetic applications including electro-osmosis, free-flow electrophoresis and isotachophoresis. These electrokinetic actuators can also serve as mechanical sensors. Because the patches are both conductive and flexible, there is a two-way conductivity-force interaction between the channel sidewall and the fluid pressure. Using this electro-mechanical feature, we also show how devices capable of performing on-chip electrophoresis and electro-osmosis can be used to measure the pressure gradient and flow rate within the same channel. Therefore, this conductive side-wall platform offers a useful means of performing electrokinetic actuation and electro-mechanical detection using the same design.

460003 Electroosmotic Flow in a Suspended Liquid Film

Abstract:

This paper reports on the electroosmotic flow profiles in a suspended liquid film stabilized with both cationic and anionic surfactants. Experimental velocity profiles were measured using confocal micro-PIV (micron- resolution Particle Image Velocimetry). It was found that the resultant flow field mainly depends on the type of the surfactant used as the shape of the film and the zeta potentials at the gas-liquid and solid-liquid interfaces are affected by the characteristics of the surfactant molecules. A computational model was also developed to predict the flow field in this system and was verified using the experimental measurements. Foam consists of large number of liquid lamella that gives rise to processes that depends on surface phenomena and application of an electric field leads to interesting flow patterns within the foam structure that could lead to new applications. In recent years, electric field interactions with free suspended films/soap films have gained significant interest. Recent work includes liquid film motors, [1] and reverse drainage of a cylindrical foam column, [2]. The main purpose of this study is to determine the flow field within a suspended foam film as a result of electro kinetic phenomena and to investigate the effect of surfactant type on the flow features with the view of finding new applications for micro/nano fluidics. A liquid film holder is fabricated by connecting two clear borosilicate glass rods with two platinized titanium rod. A liquid film was suspended in the frame by pipetting a small volume of the test solution made of deionized water, glycerol, phosphate buffer and surfactant and carefully dispensing it to the cavity of the holder to make a thin film. The liquid film was in direct contact with the electrode and the glass rods and takes the shape of a liquid lamella observed in foam. By applying an electric field of 1250 V/m across the liquid film, a flow field was established using electroosmotic flow. The resultant flow field within the cell was measure by confocal micro-PIV (micron- resolution Particle Image Velocimetry) using 1 micron polystyrene latex (PSL) particles. In order to account for electrophoresis of the flow visualization particles, electrophoretic mobility of the particles in the test solution were measured using Malvern Zetasizer Nano-ZS.

References

1. Simulation of Liquid Film Motor: A Charge Induction Mechanism Nasiri, M.; Shirsavar, R.; Saghaei, T.; Ramos, a.. Microfluid. Nanofluidics, 19 (1), 133â€"139, 2015,

2. Bonhomme, O.; Liot, O.; Biance, A.-L.; Bocquet, L. Soft Nanofluidic Transport in a Soap Film. Phys. Rev. Lett., 110 (5), 054502.2013,

461857 Selectivity Enhancements in Gel-Based DNA-Nanoparticle Assays By Membrane-Induced Isotachophoresis: Thermodynamics Versus Kinetics

Abstract:

Selectivity improvements in microfluidic chips for nucleic acid sensing remain challenging because of both thermodynamic and kinetic limitations. The thermodynamically based dissociation constants of DNA largely determine an assayâ€™s ability to discriminate between matched (target) and unmatched (nontarget) sequences. Furthermore, techniques which improve limits of detection and shorten assay times by altering thermodynamic equilibrium, such as isotachophoretic preconcentration, do little to enhance the selectivity because these improvements affect targets and nontargets equally. Introducing external forces to facilitate hydrodynamic or electrical shear may increase selectivity for assays with surface-bound probes. However, elimination of nontarget signals are also met with substantial losses in target signals thereby leading to poor limits of detection. To overcome both the thermodynamic and kinetic limitations of current DNA sensors, we introduce a platform based on depletion isotachophoresis in agarose gel generated by an ion-selective membrane. The ion-selective membrane permits us to capture target molecules with probe-functionalized nanoparticles, concentrate them, and lower the effective dissociation by more than four orders of magnitude. The membrane also generates a depletion zone to create high electrical shear forces which enable irreversible dehybridization of the nontargets. By varying the applied voltage during the depletion step, we enhance the selectivity beyond the thermodynamic limit by five-fold for nontargets with only two base mismatches while minimizing any loss in the targetsâ€™ signal. We also explore the effect of mismatch location on selectivity and demonstrate the importance the dehybridization rate constant in determining the magnitude of the selectivity.

473464 Simulation of Near-Size-Independent iDEP Separation Using Multiple Electric Fields

Abstract:

Dielectrophoresis (DEP) is a non-linear electrokinetic force that depends on the magnitude of gradients in the squared electric field and differences between electrical properties of a particle and its surrounding media. For a particle immersed in a semi-infinite media with a non-uniform electric field that is well-approximated by a truncated multipole expansion (i.e., not *too* nonlinear), the DEP force is:

$$\langle \vec{F}_{DEP} \rangle = \pi \epsilon_m a^3 \Re\left[\tilde{f}_{CM}\right] \nabla (\vec{E} \cdot \vec{E}) \qquad (1)$$

where a is the particle radius, $\Re[\cdots]$ is the ÒrealÓ operator, and $\tilde{f}_{CM}$ is the so-called complex Clausius-Mossotti factor which is dependent on the electrical properties of a particle and its surrounding fluid. Owing to this dependence, DEP-based techniques have shown great potential to isolate, separate, enrich, or otherwise manipulate biological particles (e.g., cells, vesicles, DNA). Sensitivity to particle size arises in two locations: the Clausius-Mossotti factor (for particles with complex internal structures) and in the preceding a^3 term. The cubic dependence on particle radius means that biological variability in size will likely dominate any subtler Ð through much more relevant and interesting Ð electrical differences indicative of a particular phenotype or composition of interest.

We have designed a new dielectrophoresis separation that can reduce this primary dependence on particle size. The technique combines multiple DEP fields at different frequencies with an electrically insulating constriction in channel depth. When an electric potential is applied across an insulating constriction, the electric field is concentrated, forming a local gradient in the electric field where the channel cross-sectional area changes. This gradient drives DEP forces and has been used to separate particles in a number of different systems and is typically referred to as ÒinsulatorÓ-based or ÒiDEPÓ. The device layout is shown in Figure 1A and composed of a wide straight microchannel with an angled constriction across its width. A pressure driven flow is used to carry particles through the device. Electrodes are embedded along the sidewalls on either side of the constriction, and are energized with different electric potentials at different frequencies. The two field frequencies are chosen based on analytical and empirical data: one where the DEP response is sensitive to changes in parameters of interest (e.g., increasing cytoplasmic permittivity) and one that is opposite and insensitive to changes in parameters of interest. The latter is typically low frequency such that $\Re[\tilde{f}_{CM}]$ depends on membrane conductivity and drives nDEP; the former is determined by multishell modeling or empirical data. The ÒinsensitiveÓ or nDEP field is applied uniformly along the sidewall and exerts a uniform nDEP force that exceeds the fluid drag force. The resulting particle motion is parallel to the constriction and at an angle to the direction of pressure driven flow. The ÒsensitiveÓ or pDEP field is applied to the other electrode and Ð critically Ð *increases along electrode length*. The separation depends on the sum of three forces: the uniform nDEP force, the increasing pDEP force, and the fluid drag force. As the pDEP force increases, the sum of the pDEP and nDEP forces decreases until the nDEP force no longer exceeds the drag force and the cell is transported with the fluid out of the device. By driving the magnitude and spatial variation in nDEP and pDEP forces, it is possible to reduce the separations dependence on the drag force relative to the DEP force. In preliminary work, we have shown that, using a multishell model, it is possible to use this technique to distinguish bacterial cells with membrane permittivities differing only by a factor of two (Figure 1B).

Key benefits of this design are: continuous flow separation, variable sensitivity based on the chosen electric potentials, and reduced sensitivity to particle size (Figure 4). One feature of the proposed technique is its ability to tune separation sensitivity by changing the applied electric potentials, the fluid volumetric flow rate, and the frequency of applied signals. Compared to other systems, this device can operate on a wider range of cell samples and operates in continuous flow.

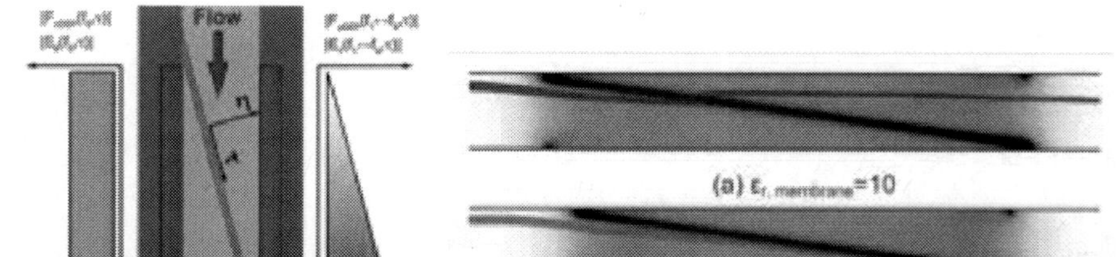

Figure 1: *(A) Schematic of applied electric fields and resulting F_{DEP}. Flow from top to bottom, electric fields are applied on the left (nDEP) and right (pDEP) electrodes (red). Electric field gradients are generated by a diagonal constriction in channel depth. nDEP field (orange) is constant and pDEP field (green) increases along channel; particles are deflected along constriction by nDEP until $|F_{pDEP}| = |F_{nDEP}|$ and fluid drag dominates. (B. a, top) 2D simulations of particle trajectories using the electrode scheme in (A). Flow is from left to right and colortable corresponds to total electric field magnitude (black is highest over the*

468247 Non-Optical Biomolecular Detection in Human Serum Using Interfacial Electrokinetic Transduction

Abstract:

Development of rapid, sensitive and portable detection systems are important for effective detection of diseases in developing countries, biowarfare/anti-terrorism applications, environmental monitoring, and for basic biological research. Biosensors are analytical devices that detects an analyze of interest by combining biological sensing elements with physiochemical transducers. While current biosensing methods are capable of sensitively detecting a variety of target analytes in a laboratory setting, there are inherent difficulties in developing low-cost portable biosensors for point-of-care diagnostics using traditional optical, mass, or electroanalytical-based signal transducers. It is therefore important to develop new biosensing transducer elements for recognizing binding events at low cost and in portable environments. This approach utilizes a novel electrical method that combines a laminar liquid-liquid interact with alternating current (ac) electrokinetics. A bioreaction occurs at the co-flowing liquid interface while an ac electric field is simultaneously applied, displacing the interface across the flow channel. The magnitude and direction of displacement is directly related to both the applied frequency and interfacial electrical properties. Biosensor detection is based on interfacial electrokinetic transduction; binding changes the interfacial electrical properties, which is electrically traduced and detected by monitoring the frequency response of the liquid interface. In this manner, the liquid interface serves as both the substrate for specific binding, and transducer for biomolecular recognition. Furthermore, by implementing electrical impedance spectroscopy (EIS), the interfacial position can be monitored non-optically. Utilizing EIS, the biosensor becomes label-free and non-optical; the magnitude of impedance, which can be monitored on a computer screen, becomes the readable signal. Here, a non-optical, label-free biosensor, which can detect bimolecular reactions in human serum is created by pairing interfacial electrokinetic transduction and electrical impedance spectroscopy.

469927 A Membraneless Microfluidic Architecture for Continuous Separation of Particles and Cells

Abstract:

We have developed a microfluidic filtration architecture based on an embedded weir-like barrier separating two lanes with unequal depths. The barrier is oriented parallel to the flow direction and extends along the entire centerline length of the microchannel, making it possible to process large volumes with no clogging. Here we report coordinated experiments and computational simulations aimed at quantifying the interplay between longitudinal and lateral pressure distributions within the microchannel that govern the achievable separation efficiency in this system. These new studies reveal that the effect of unequal microchannel depths on each side of the central barrier can be coupled with a transverse centrifugal Dean flow introduced along a curved flow path segment, enabling greatly enhanced separation and enrichment to be achieved across a broad range of flow rates. A 3D computational fluid dynamics model was developed to capture the features of the secondary flow field, while a coupled discrete element modeling algorithm tracked trajectories of individual species under the influence of all relevant hydrodynamic forces and torques. This analysis enables a universal correlation to be established linking all device design parameters (cross-sectional area profile, length of microchannel, flow rate), permitting rational design for efficient size-based filtration of desired cellular and extra-cellular blood components. We broadly apply these fundamental insights to demonstrate high-throughput separation of species ranging from micron-sized particles, to cancer cells, to extracellular materials. We also introduce a novel micro-fabrication method that can be readily scaled to enable mass production of filtration chips suitable for a host of cell enrichment and screening applications.

468157 Detecting Autologous Blood Transfusions Using Dielectrophoretic Spectroscopy

Abstract:

The desire to increase performance in professional athletes has driven some to turn to artificial methods to achieve this end. Autologous blood transfusions – the collection and subsequent reinfusion of a person's own blood – currently lack a direct test for detection, and has for that reason become a popular "banned method" in performance enhancement. The Athlete's Biological Passport is meant to monitor an athlete's long-term trends in numerous biomarkers, and use abnormal variations in the markers to potentially identify blood-doping. It does provide an indirect means of identifying autologous blood doping, but there still lacks a direct detection method.

Using a combination of membrane protein cross-linking and dielectrophoretic spectroscopy, we demonstrate the ability to directly detect the presence of 5% reinfused red blood cells (RBCs) in fresh blood. This corresponds to one transfusion unit of whole blood.

Red blood cells aging in storage undergo biochemical changes that normal RBCs in circulation do not. This includes changes in cell membrane composition and cell morphology. When stored ex vivo, RBCs begin to lose the number of membrane-bound proteins. This physiological change offers less surface proteins available for cross-linking, resulting in a different dielectrophoretic spectra for stored cells compared to that of fresh RBCs. We use differences in these spectra to discern storage-related changes to blood cell morphology and composition, as a means to directly detect autologous blood doping in athletes.

470451 Long-Read DNA Separations Using Micelle-Elfse in Microchip Electrophoresis

Abstract:

Our group has developed a gel-free method (â€œmicelle-ELFSEâ€) that provides rapid separation of end-alkylated DNA fragments in running buffers containing 1-3 vol% of nonionic wormlike surfactant micelles. These micelles transiently attach to the alkyl group and serve as a drag-tag so that the free-solution mobility of the fragments in length dependent. Micelles undergo rapid fluctuations in size so that the drag imparted to each fragment in the sample is highly uniform. We have recently showed that the method can resolve Sanger sequencing products over 700 bases in length, with single-base resolution, in less than 15 mins using standard bench-top capillary electrophoresis equipment. Micelle-ELFSE is simple to implement in bench-top capillary electrophoresis instrumentation, uses low-cost materials, and the buffers used do not clog in the presence of contaminants. Micelle size can be tuned to fit a given separation by temperature or formulation.

In this presentation, we will discuss our efforts to apply micelle-ELFSE to much longer DNA fragments, well over 1000 bases in length. Kilobase DNA separations are required for many applications in biology and medicine, including gene mapping, pathogen identification, and processing of chromosomal DNA digests in next-generation sequencing technologies. In DC mode, gel-based DNA electrophoresis fails for fragments above 15,000 bases in length due to biased reptation effects. Pulsed-field electrophoresis methods are capable of separating fragments above 15,000 bases, but with runtimes of several days. An alternative is the use of microfabricated post arrays and entropic traps, but these are expensive to implement and have low throughput. Micelle-ELFSE methods are not expected to suffer from biased reptation limits as the wormlike micelles used have large effective pore sizes and short micelle lifetimes. To test this concept, we created a ladder with fragments 1000, 2000, 3000, 4000, 6000, and 10,000 bases in length and were able to separate it, with baseline resolution, in 3 minutes using bench-top capillary electrophoresis instrumentation. The success of the method relies on an apparent DNA-tag segregation that causes the peak spacing to be larger than predicted by theories that assume the tag and DNA to be hydrodynamically indistinct. Additionally, we have found that use of very long wormlike micelles, which form an entangled network capable of sieving unalkylated DNA, does not impact the effectiveness of micelle-ELFSE. Access to very large micelle drag-tags greatly extends the lengths of DNA that can be resolved.

We have since created ladders exceeding 30,000 bases in length and will optimize large-micelle buffers for their separation for presentation in this talk. The ultimate goal is to determine the fastest runtimes and DNA lengths that can be resolved using micelle-ELFSE. We will also report an observed buffer dependence of the DNA-tag segregation that suggests new mechanisms for segregation. Finally, we will discuss efforts to perform these separations in microfluidic devices, where snap-shot imaging of the full microchannel can further reduce runtime.

464410 A Tunable Ionic Transistor/Diode Molecular Sensor with Adjustable Sensitivity and Dynamic Range

Abstract:

We report a new control ionic circuit for a membrane biosensor developed in our lab [Slouka et al, Annu Rev Anal Chem, 7, 317 (2014)]. The ionic circuit is designed such that the sensitivity and dynamic range of the membrane sensor can be tuned by a gating voltage, both over several orders of magnitude.

The principle behind the membrane sensor is based on ion depletion and charge polarization phenomena that occur at one ion-selective membrane surface when an electric field is applied across it, thus producing the well-known limiting and overlimiting currents. These two latter regimes of the I-V curve occur at much higher voltages (1-10 V) than the traditional electrochemical sensors that function at the redox potentials, thus offering much better signal-to-noise ratio. The biomolecular targets, captured by probes functionalized onto the membrane surface, are counterions. Their immobility after hybridization with the probes impedes the depletion action and produces an effective bipolar membrane, thus significantly delay the development of electroconvection vortices or enhance water splitting, the two key phenomena responsible for the overlimiting current [Slouka et al, Langmuir, 29, 8275-8283 (2013)].

The sensitivity of the membrane sensor is determined by the lowest fraction of hybridized probes that can be detected by the voltage shift in the overlimiting current regime. Scaling down the membrane area can significantly lower the detection limit but may also limit the dynamic range. We hence design an ionic circuit that can amplify this voltage drop with proper impedance matching. A key feature of the signal amplifying ionic circuit component is an ionic transistor, which allows a small normal gating voltage signal to control a large draining ionic cross current [Sun et al, LabChip,16, 1171-1177 (2016)]. The amplification in the voltage has been shown to exceed 10-fold. We use a bipolar diode membrane design with different external ionic circuits such that the amplification factor can be tuned by the gating voltage, thus effectively shifting both the sensitivity and dynamic range of the sensor. We are testing this ionic circuit design on nucleic acid sensing and expect to achieve a dynamic range of fM to mM with the same membrane chip.

470348 Investigation of the Effect of Electroporation on Chemotherapeutics Delivery into a Tumor

Abstract:

Due to biological barriers in a tumor microenvironment, drug delivery into the tumor will experience challenges. For example the tumor microenvironment is very heterogeneous due to angiogenesis, creating significant challenges for effective drug delivery. Application of an electric field has the potential to overcome some of these barriers. Electric field can enhanced drug transport by increasing the cell permeability and increasing the tumor drug uptake rate. Applying electric field in a cancer cell leads to three major electrokinetic phenomena namely, electrophoresis (particle/ion movement), electroosmosis (ion fluid movement) and electroporation. Electroporation phenomena facilitates the introduction of chemotherapeutics particles into the cancer cell by increasing the permeability in the blood vessel membrane. Computational models have been developed to study some of these phenomena [1-4]. However, to our knowledge, a computational model that simultaneously considers all three phenomena is missing in the literature. In Moarefian et al. [5] a mathematical model was developed and subsequent simulations have suggested that applying electric field will enhance chemotherapeutics delivery. The authors did not consider electroporation effect on organelles in the cancer cell and how the electroporation could increase drug diffusion. Finally they ignored the electroosmosis effect in the intercellular matrix around the cancer cell and in the blood vessel. In this presentation we will discuss a new model that relaxes the assumptions discussed above. The hypothesis is that the model would lead to a significantly more realistic prediction of the optimum electric field that would result in the maximum fraction of cell killed. A platform to test and validate has not been performed by Moarefian et al.[5]. Here we proposed the preliminary design of microfluidic device to study the effect electroporation on drug delivery.

1. Pascal, J., et al., Mechanistic patient-specific predictive correlation of tumor drug response with microenvironment and perfusion measurements. Proceedings of the National Academy of Sciences, 2013. 110(35): p. 14266-14271.
2. Edd, J.F. and R.V. Davalos, Mathematical modeling of irreversible electroporation for treatment planning.

Technology in cancer research & treatment, 2007. 6(4): p. 275-286.
3. Krassowska, W. and P.D. Filev, Modeling Electroporation in a Single Cell. Biophysical Journal, 2007. 92(2): p. 404-417.
4. Sersa, G., et al., Vascular disrupting action of electroporation and electrochemotherapy with bleomycin in murine sarcoma. British journal of cancer, 2008. 98(2): p. 388-398.
5. Moarefian, M. and J.A. Pascal, Fundamental mathematical model shows that applied electrical field enhances chemotherapy delivery to tumors. Mathematical Biosciences, 2016. 272: p. 1-5.

472497 A High-Throughput Platform for Electrotransformation of *E. coli*

Abstract:

Synthetic biology holds the potential for solving many pressing challenges for mankind and our planet. However, one technical challenge in any area that relies upon genetic manipulation of cells is the low throughput encountered in many forms of genetic transformation. We developed a new continuous flow microfluidic system for bacterial genetic electrotransformation using pulsed electric fields. The proposed microfluidic platform has the potential to outperform the state-of-the-art cuvette electroporation by several orders of magnitude, based on its increased throughput.

Mammalian cell electroporation in microfluidic devices has demonstrated significantly improved transfection efficiency and higher cell viability in comparison to bulk, cuvette electroporation. The increase in transfection efficiency in flow-through microfluidic devices uses a fraction of the experimental sample and lower voltages, maintaining high transfection efficiency and high cell viability. Despite substantial advances in transfection of mammalian cells in microfluidic devices, an electrotransformation platform for high-throughput processing of bacteria without requiring physical microbe modifications (e.g. magnetic beads and/or oil droplets) has not been developed. A microfluidic platform for continuous electrotransformation of bacteria (nominal size ~1 μm) that achieves maximum pulsed electric field ~15 kV/cm is presented here. The microfluidic channel employs a non-uniform constriction to generate high electric fields to induce bacterial electrotransformation without generating lethal Joule heating. In this device, cells experience a time-dependent electric field that is hydrodynamically controlled, allowing cells to experience conditions that would be challenging to achieve with standard electronics.

Escherichia coli DH5a in exponential phase (MIT, Boyer Lab) was used to demonstrate the high-throughput electrotransformation platform. Pulsed electric fields (2.5 kV and 5-ms square pulses with a 20 % duty cycle) were delivered in the presence of DNA-coding (Parts Registry K176011) for ampicillin resistance and green fluorescent protein (GFP) at a final DNA concentration of $C_{DNA} = 1$ ng/μL. The electroporation buffer consisted of 10 % (v/v)

glycerol supplemented with 0.05 % (v/v) Tween 20 in order to mitigate cell-to-cell agglomeration. Each experimental sample (100 μL) was driven at 0 μL/min (1.97×10^8 CFU/μgDNA), 250 μL/min (2.26×10^8 CFU/μgDNA), or 500 μL/min (2.97×10^8 CFU/μgDNA) and resulted in a residence time (pulse duration) within the constriction < 5 ms. We achieved high transformation efficiencies with a throughput increase of up to three orders of magnitude compared to the state-of-the-art cuvette electroporation. This work facilitates high throughput electrotransformation of microorganisms, accelerating development of genetically engineered microbes for important industrial applications. Electrotransformation of microbes is an essential part of many scientific fields including the study of pathogenic microbes, metabolic engineering, synthetic biology, and human microbiome for therapeutic applications. With further development, a flow through microfluidic electrotransformation platform will be realized for discovery of electroporation conditions for difficult-to-transfect or intractable microbes. We envision that the proposed microfluidic technology will reduce electrotransformation techniques that can take months down to a couple hours.

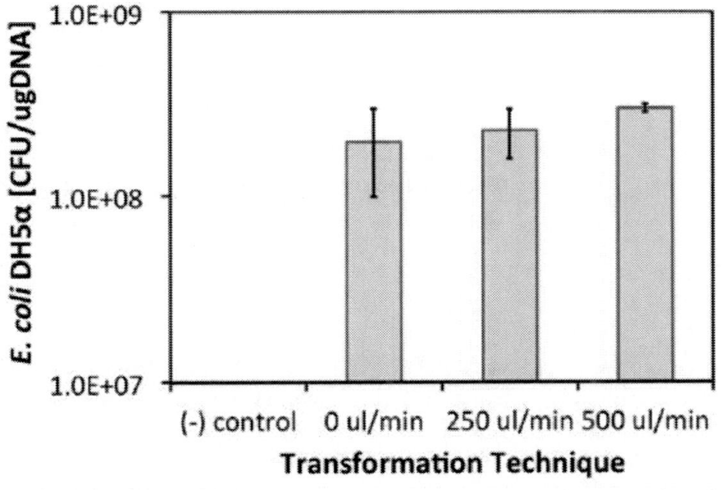

474811 Microfluidic Cell Sorting and Microphysiological Circulation: From Liquid Biopsy to Vascularized Micro Tissue

Abstract:

Over the last 20 years, many microfluidic technologies have been developed for sample preparation at the cellular scale to enable on-chip flow cytometry, cell separation, cell enrichment, and single cell trapping/release. Examples of these microfluidic sorting technologies include dielectrophoresis (DEP), acoustophoresis, acoustic streaming, magnetophoresis, immunomagnetic, magnetohydrodynamic, and inertial microfluidics. The basic principle is to exert forces at controlled levels above the laminar flow-induced viscous hydrodynamic forces in order to separate the targeted cells of interest. Liquid biopsy has become a promising technology to isolate and target rare cells such as circulating tumor cells (CTCs) in body fluids thanks to many of these microfluidic cell sorting techniques. This advent of microfluidic liquid biopsy provides an *in vitro* snap shot into the patientâ€™s physiological status via the *in vivo*circulation that enables one to monitor disease state and progression for diagnosis and prognosis.

Recently there has been a surge in the development of microphysiological systems and organ-on-a-chip for drug screening and regenerative medicine. Over the years, drug screening has mostly been carried out on 2D monolayers in microwell plates and the drugs screened are not delivered through blood vessels as they are for *in vivo* treatments. Through the advancement of microfluidics technologies, our team has enabled the automation of biological fluids delivery through physiological vasculature networks that mimic the physiological circulation of the human body. The critical bottleneck is to engineer the microenvironment for the formation of 3D tissues and organs and to also pump and perfuse the tissue vascular network for on-chip *in vitro* microcirculation. Microfluidics play an important role in both the above-mentioned *in vivo* liquid biopsy and *in vitro* physiological circulation platforms. These two technologies will go hand-in-hand to connect *in vitro* screening to *in vivo*treatment with tremendous potential towards the realization of personalized medicine.

474814 Engineering Paper Microfluidic Sensors for Point-of-Care Applications in Low-Resource Settings

Abstract:

High-sensitivity point-of-care diagnostic assays that are rapid, easy to use, and low cost are needed for use in low-resource settings. The conventional lateral flow test is the standard bioassay format for point-of-care applications in these settings due to its rapid time to result, ease of use, and low cost. However, for many analytes, conventional lateral flow tests lack the sensitivity to have clinical utility. The rapidly growing field of "paper" microfluidics has the potential to address this issue. In particular, we are developing paper-based devices that can automatically perform multi-step sample processing for high-sensitivity detection in a format that is appropriate for use in low-resource settings. Key to the successful operation of these devices is the development of a "paper microfluidics toolbox" for controlling flow and metering multiple small volumes of fluids within devices. This presentation will highlight tools development in the context of applications for infectious disease diagnosis (e.g., malaria and influenza) and therapy monitoring (e.g., for phenylketonuria treatment) in low-resource settings.

474815 Label-Free Magnetic Additive Biomanufacturing Technologies to Isolate and Sort Circulating Tumor Cells and Microemboli

Abstract:

Micro- and nano-scale technologies can have a significant impact on medicine and biology in the areas of cell manipulation, diagnostics and monitoring. At the convergence of these new technologies and biology, we research for enabling solutions to the real world problems at the clinic. Emerging nano-scale and microfluidic technologies integrated with biology offer innovative possibilities for creating intelligent, mobile medical lab-chip devices that could transform diagnostics and monitoring, tissue engineering and regenerative medicine. In this talk, we will present an overview of our laboratory's work in these areas focussed on applications in magnetic levitation methods for assembling cells and label free sorting of rare cells from whole blood. Cells consist of micro- and nano-scale components and materials that contribute to their fundamental magnetic and density signatures. Previous studies have claimed that magnetic levitation can only be used to measure density signatures of nonliving materials. Here, we demonstrate that both eukaryotic and prokaryotic cells can be levitated and that each cell has a unique levitation profile. Furthermore, our levitation platform uniquely enables ultrasensitive density

measurements, imaging, and profiling of cells in real-time at single-cell resolution. This method has broad applications, such as the label-free identification and sorting of CTCs and CTM with broad applications in drug screening in personalized medicine.

474816 The Cellular Circadian Clock Drives Daily Rhythms of Ion Transport

Abstract:

Circadian clocks are fundamental to the biology of most organisms, coordinating physiology and behaviours (such as sleep) to resonate with the environmental cycle of day and night, through complex networks of clock-controlled genes. Circadian disruption in humans, as occurs during shift work, is strongly associated with a range of chronic diseases e.g. type II diabetes, various cancers. A fundamental knowledge gap exists, however, between clock gene expression cycles and the biochemical mechanisms that ultimately imparts circadian regulation to the activity of individual cells. Using a wide range of model organisms, including human, fungal and algal cells, as well as techniques such as dielectrophoresis and inductively-coupled plasma mass spectrometry, we have observed a daily rhythm in the import/export of K^+ and Mg^{2+}ions over the cellâ€™s plasma membrane. The striking evolutionary conservation of these daily rhythms of ion transport, combined with their functional consequences for cellular energy balance, suggests a fundamental yet unexplored feature of eukaryotic circadian timekeeping that has persisted from a common ancestor living >1.5 billion years ago.

474817 Inspection of Solubilization Loci of Functional Series into Micelle Compartments as a Guide to Improve MEKC Selectivity

Abstract:

Micelles can be viewed as entities composed of numerous compartments of distinct hydrophobicities: inner core, palisade, solvation layer, and surface are readily defined in most micellar systems. We have proposed that MEKC selectivity can be rationalized by considering the details of the micellar phase and the manner by which solutes solubilize into the different micelle compartments. By knowing the major locus a solute set occupies into the micelle it is possible to promote changes in that specific locus and therefore improve the solute set resolution.

In our research group, quantitative structure-retention relationships (QSRR) and molecular dynamics experiments with a series of model compounds such as terpenes, phenolic compounds (phenolic acid and flavonoids, anthocyanins included), aromatic amines, and steroids have been used to provide insights on solute partition mechanisms and to define solubilization locus into SDS and LRO micelles.

As a preliminary assessment of solute localization we have proposed the inspection of log k (k being the retention factor in MEKC separations) versus the solute McGown volume plots. We were able to demonstrate for a large number of homologous and functional series that the slope of such plots provides a rough-tuning of solute micelle localization.

Thermodynamic parameters (enthalpy and entropy of partitioning) can be further estimated using van't Hoff plots (log P versus 1/T) for different surfactant systems. When linear van't Hoff plots are obtained we have proposed that solutes occupy single micelle compartments, once no variation of phase transfer DH and DS with temperature are observed. A curved van't Hoff plot suggests that the solute spends its residence time partially between two layers (core/solvation layer, core/palisade or palisade/solvation layer) depending on

the kind of surfactant considered. Therefore, linearity of van'Hoff plots in conjunction with the examination of the solute structural features and availability of the oxygen atoms to the solvent guide the assignment of solute location for many solute sets.

Experimentally obtained transfer DH and DS can be correlated with molecular descriptors to provide further insights into solute localization. Descriptors based on combinations of atomic charges and solvent accessible surface areas, solute molecular volumes, among others have been proven the most effective.

Our overall results revealed that the solubilization loci of terpenes is the micelle inner core, steroids occupy the internal interface of the solvation shell and flavonoids, anthocyanins and amines anchor at the surface. Therefore, to enhance the separation of flavonoids, anthocyanins and amines the basicity of surface sulfate in SDS micelles for instance must be changed by introducing spacers, such as co-surfactants and anchored solvents such as pentanol, etc. For terpenes, the inner core must be changed which can be achieved by selecting surfactant tails of different chain lengths. And finally, to optimize separations of steroids, the use of mixed micelles such as SDS/non ionic surfactants might be attempted.

455897 Colloidal Dielectric Forces within an Electric Curtain

Abstract:

The electric curtain is a platform first developed by Masuda [1] to lift and transport charged particles, typically in air, and relevant forces have typically been associated with the interaction of the applied electric field with *charged* particles. However, recent work by our group has demonstrated, contrary to traditional electric curtain theory, that *dielectric* phenomena is significant for particle manipulation. Preliminary results demonstrate that the magnitude and direction of a manipulated particle is a function of the applied AC frequency.

An electric curtain consists of a series of parallel coplanar electrodes that generate a standing wave (SW) or travelling wave (TW) AC electric field to lift particles from the platform surface and, for the TW case, simultaneously translate them away. This multiphysical system is rich in electrokinetic and mechanical physics; particle motion is governed by particle electrokinetic properties, particle mass, field properties, and medium properties [2-5]. The assumed operation was the manipulation of *charged particles* via Coulombic forces; particles are either inherently charged or they obtain a net charge via tribocharging or absorption of gaseous ions. However, dielectric phenomena are typically overlooked.

Our simple electric curtain system consists of a printed circuit board (ExpressPCB) with four sets of repeating interdigitated electrodes (0.020 inch width and spacing) where a SW (0°-180°-0°-180°) or TW (0°-90°-180°-270°) AC signal was applied with a custom four-channel step-up amplifier. The electrodes were covered with a dielectric of polypropylene tape (ε_r = 2.2-2.36). For preliminary tests Martian dust simulant (JSC MARS-1A, Orbital Technologies Corp.) was used with a particle diameter of 1 mm or smaller with over 50% of the particles (by weight) had diameters greater than 0.25 mm.

The electrokinetic-induced deflection of a suspended AFM cantilever functionalized with a 20 μm diameter borosilicate glass sphere was measured. The cantilever was suspended 34 μm away from the surface and

aligned parallel with the electrodes. Its sinusoidal deflection oscillated at *twice* the applied AC frequency, typical of a dielectric response.

A sample of solid soda lime glass beads (180-212 μm, SLGMS-2.5, Cospheric) coated the surface of the electric curtain platform prior to the activation of a SW or TW field. A high speed camera acquired images showing the interactions and positions of particles relative to the electrode. At 50 Hz particles were repelled for both SW and TW fields. For both scenarios particles experienced either low velocity oscillatory "rolling" and/or rapid projectile motion. Additional experiments showed that the repelled directionality of the particles is a function of the applied AC signal and particle type.

Further understanding of related dielectric phenomena is needed to bridge the disconnect between experimental observations and electrokinetic theory. Future results will advance applications based on such discoveries, including air-based dielectrophoresis systems for selective particle sorting or macroscale "smart filter" applications for particle detection.

References:

1. "Confinement and transportation of charged aerosol clouds via electric curtain," S. Masuda, K. Fujibayashi, K. Ishida, H. Inada, Electrical Engineering in Japan, **92**, 43 (1972).
2. "Study of dust removal by standing-wave electric curtain for application to solar cells on Mars," P. Atten, H.L. Pang, J.-L. Reboud, IEEE Transactions on Industry Applications, **45**, 75 (2009).
3. "Nonlinear dynamics of particles excited by an electric curtain," O.D. Myers, J. Wu, J.S. Marshall, Journal of Applied Physics, **114**, 154907, (2013)
4. "Electrodynamic dust shields on the international space station: exposure to the space environment," C.I. Calle, P.J. Mackey, M.D. Hogue, M.R. Johansen, H. Yim, P.D. Delaune, J.S. Clements, Journal of Electrostatics, **71**, 257 (2013).
5. "Some techniques on electrostatic separation of particle size utilizing electrostatic traveling-wave field," H. Kawamoto, Journal of Electrostatics, **66**, 220 (2008).

471543 In-Situ Characterization of Electrophoretic Deposition Using the Quartz Crystal Microbalance

Abstract:

Electrophoretic deposition (EPD) is a process in which colloids in a liquid suspension are deposited onto an electrode using an electric field. EPD allows for fast deposition over complex shapes, is easily scalable to large areas, and can be used with a diverse material set. Recent work has demonstrated that EPD can be used as a three-dimensional printing technique [1], renewing an interest in developing more quantitative deposition models for nanocrystal colloids. Since there are few studies that use *in situ* methods to quantify particle deposition, the mechanisms of electrophoretic deposition are explored traditionally through analytic models that seek to capture EPD kinetics. In our work, we use a Quartz Crystal Microbalance (QCM) to determine particle deposition in real-time. The QCM is a piezoelectric device that dynamically measures the change in resonant frequency due to particles depositing onto the face of the QCM, which acts as the electrode. We then use the frequency shift to extract deposited mass as a function of time, which enables exploration of several aspects of deposition. With the QCM, we measure the minimum voltage needed for particles to begin depositing, the deposition rate, and the dissolution rate (for unstable colloid films). A comparison with measured current response allows us to correlate mass transfer with the charge transfer of the particles. By varying deposition parameters such as particle concentration, deposition time, and applied electric field, we develop datasets that compares different colloids and solvents. Here, we present our experimental techniques and results from parameter studies compared with analytic models of our system.

This work was supported by the Lawrence Livermore National Laboratory Directed Research and Development Program, 16-ERD-033 and performed under the auspices of the U.S. Department of Energy by Lawrence Livermore National Laboratory under Contract DE-AC52-07NA27344. Work at the Molecular Foundry was supported by the Office of Science, Office of Basic Energy Sciences, of the U.S. Department of Energy under Contract No. DE-AC02-05CH11231.

[1] Pascall, et al. Advanced Materials. 2014.

467572 Electrokinetics of Heterogeneous Ion-Exchange Membranes

Abstract:

Ion selective systems [1] such as nanochannels, ion exchange particles or ion exchange membranes exhibit very interesting electrokinetic behavior that is increasingly used in microfluidic systems for applications related to point-of-care diagnostics such as sample pretreatment, biomolecule preconcentration or sensing [2]. All these systems are characterized with a nonlinear current voltage curve which shows three distinctive regions denoted as: (i) underlimiting, (ii) limiting and (iii) overlimiting. We showed that some heterogeneous ion exchange membranes can interact with specific biomolecules possessing the ability to change the current voltage curves in the limiting and overlimiting region above all. Based on this membrane response to the presence of the biomolecules, we developed a sensor for specific detection of single stranded nucleic acids. Interestingly, the response of ion exchange particles which are functional parts of heterogeneous membranes, to the same molecules did not result in quantitatively same behavior. This observation raised a question about particular features of the membranes, especially those related to the structure of the membrane, that render the membrane sensitive to the biomolecules.

To investigate the effect of membrane structure on the exhibited behavior, we developed a system that integrates a small piece of ion exchange membranes and allows to (i) reconstruct swollen heterogeneous membranes by means of micro-computed tomography and (ii) study their electrochemical and electrokinetic behavior when connected in DC field. In this contribution, we will show the effect of ionic strength of water solutions on swelling and shrinkage of membranes and associated irreversible changes in the membrane structures. This will be followed by experimental analysis of the electrokinetic behavior of these membranes when connected in overlimiting region and discussion about the role of structure in appearance of this region and thus sensitivity to the presence of biomolecules.

[1] Slouka, Z., Senapati, S., Chang, H.C., Microfluidic systems with ion-selective membranes, *Annual Review Analytical Chemistry*, 2014; 7:317-335

[2] Senapati, S., Slouka, Z., Shah, S. et al. An ion-exchange nanomembrane sensor for detection of nucleic acids using a surface charge inversion phenomenon, *Biosensor&Bioelectronics*, 2014, 60:92-100

469877 Mesoscale Particle-Based Model of Electrophoretic Deposition

Abstract:

Electrophoretic deposition (EPD) describes the process in which field-driven particles accumulate at electrodes and assemble into layers. The majority of EPD models provide continuum level descriptions of mass deposition that inherently neglect particle-based interactions within the bulk suspension, growth front, and deposit. These inter-colloidal interactions are necessary to understand the underlying physics that govern particle rearrangement, defect formation, and the ultimate microstructure of the deposit. Here, we present and evaluate a particle-based model of colloidal suspensions that undergo electrophoretic motion[1] and deposition using an extensive set of mesoscale simulations that characterize experimentally relevant colloidal suspensions. The model accurately describes diffusive and electrophoretic motion of experimentally relevant colloidal suspensions. Since the model explicitly computes inter-colloidal interactions, it is uniquely poised to elucidate how deposition conditions influence defect structures and particle rearrangement within EPD colloidal crystals. We use the model to investigate how empirical parameters, such as electric field strength and electrolyte concentration, can be tuned in order to control the degree of colloidal ordering versus non-ordering that occurs during EPD. It is straightforward to configure the model to study how various preparations of the interface, e.g. a bare surface, a lattice of particles, an amorphous monolayer, etc., and also annealing schemes influence the deposit microstructure.

This work was performed under the auspices of the U.S. Department of Energy by Lawrence Livermore National Laboratory under Contract DE-AC52-07NA27344.

[1] Giera, B.; Zepeda-Ruiz, L. A.; Pascall, A. J.; Kuntz, J. D.; Spadaccini, C. M.; Weisgraber, T. H. Mesoscale Particle-Based Model of Electrophoresis. *Journal of the Electrochemical Society*. The Electrochemical Society 2015, pp D3030â€"D3035.

462253 Effect of Intedigitated Electrode Asymmetry on Performance of Carbon Based AC Electroosmotic Micropumps

Abstract:

The increased research interest in Alternating Current Electroosmotic (ACEO) micropumps over the past few decades is due to the many advantages ACEO pumps present compared to other approaches. These advantages include the low power consumption and the absence of movable mechanical parts, which reduces complexity and manufacturing cost. Although a few studies have been previously reported on the effects of electrode geometry on the performance of ACEO micropumps, it is still unclear how the asymmetry ratio of interdigitated electrodes affects the pumping rate of these devices. The present research investigates the effect of the asymmetry of interdigitated electrodes on the performance of carbon-based ACEO micropumps. The conventional carbon-based microelectromechanical systems (C-MEMS) technique which employs photolithographic patterning of a photoresist precursor, followed by pyrolysis to form glassy carbon structures, is utilized to fabricate interdigitated carbon microelectrode arrays.

To explore the effect of asymmetry of interdigitated microelectrodes on the overall fluid velocity, planar carbon electrode arrays with electrode width ratios of 60/20 µm, 80/20 µm, and 100/20 µm, with separation between electrode pairs of 20 µm were fabricated. Bi-distilled water (σ = 1.72 µS/cm) was used as a working fluid to measure the pumping rate of carbon ACEO micropumps. An AC voltage ranging from 2 to 20 V_{pp} at frequencies from 1 to 200 kHz was applied to induce ACEO flow in a polydimethylsiloxane microchannel 100 microns high and 500 microns wide. For all the cases, as the AC frequency decreased, the fluid velocity increased in the applied frequency range. At low amplitudes (less than 8 V_{pp}), the fluid velocity increased as the voltage increased. The fluid velocity was decreasing at mid-range amplitudes (10-14 V_{pp}), and the flow direction was reversed at high amplitudes (16 - 20 V_{pp}). The flow reversal at high amplitudes might be attributed to the dominant ACEO vortex reaching the top of the microchannel, eliminating the conveyor-belt effect thus propelling the fluid to the opposite direction. As the asymmetry of

interdigitated electrodes decreased, the fluid velocity increased for both forward and backward flows. The maximum observed fluid velocities were 28.6 µm/s, 11.44 µm/s and 7.3 µm/s in the forward direction, and 338 µm/s, 106.56 µm/s and 79.65 µm/s in the backward direction, for the interdigitated carbon microelectrodes with ratios of 60/20, 80/20, and 100/20 µm, respectively. Interestingly, the flow rate increased as the ratio of asymmetry decreased for the tested ratios. The likely explanation for this trend is that as the width of the wider electrode decreases, portion of the dominant flow over this electrode increases, thus more fluid volume is driven by this vortex. We expect that findings of this study will be useful for the design of high performance ACEO micropumps based on asymmetric interdigitated electrode arrays.

470020 A Shear-Enhanced CNT-DEP Nanosensor Platform for Ultra-Sensitive/Selective Protein Quantification with Tunable Dynamic Range: Overcoming Thermodynamic Limitations

Abstract:

Detection and quantification of low-concentration (<nM) proteins in heterogeneous media are plagued by two distinct obstacles: lack of sensitivity due to high dissociation equilibrium constants K_D of most antibody probes in the range of (μM - nM) and non-specificity due to an abundance of non-targets with similar K_D. We report a new nanoscale protein sensor platform that uses irreversible electrokinetic phenomena to overcome these thermodynamic limitations. The sensitivity is achieved with irreversible rare events, driven by a DEP (dielectrophoresis) driven Ab-Ag-Ab ELISA complex association, which can nevertheless be detected with a sensitive CNT electron tunneling conductance sensor design. The selectivity is achieved by working at a critical shear rate that is between the dissociation shear rate of the target and the non-target, such that a few kT difference in their complexification enthalpy can drastically change their dissociation rate. The CNT reporter plays several important roles in this platform: as an inert but sensitive electron tunneling reporter and as a selectivity enhancer by virtue of its large hydrodynamic viscous drag. We design our current protein ELISA platform based on our earlier CNT sensor work for bacteria sensing[1] and DNA sensing[2]. Tunable dynamic range and ultra-sensitivity are achieved here with a CNT-Ag lock-switch electron tunneling titration design.

Specific components of this new protein sensing platform include: 1. Accelerated DEP association of protein target with an Ab probe functionalized onto an electrode pair with high electric field; 2.Transportation and trapping of the second linking Ab probe functionalized CNTs with long-range DC electrophoresis and short-range AC DEP to the electrodes (lock); 3.Selective removal of assembled CNTs with non-targets by an optimized nanoshear protocol;

4.Conductance quantification of remaining bridging CNTs with target linkers (switch). Steps 1 and 2 involve rapidly driven dynamic events to prevent the captured targets from dissociating from the probe to reach the thermodynamic coverage. Optimized shear at the nm-high hydrodynamic slip length in step 3 irreversibly removes CNTs with non-targets due to the large Stokes drag of their high aspect-ratio cylindrical geometry. With only target-linked CNTs remaining, the digital conductance by electron-tunneling across the Ab-Ag-Ab complex allows us reach a detection limit of 100 molecules (100 aM). Irreversible capture and shearing also allow us to tune the dynamic range up to 100 billion (100 pM or 6 decades) by increasing the CNT number. We will also demonstrate carefully designed selectivity studies against non-targets with similar K_D that report orders of magnitude discriminating factors. The data will be analyzed with a Langmuir association theory to quantify the deviation from thermodynamic equilibrium and how much shear reduces the dissociation barrier. Concentration factor and how DEP and electrophoresis reduces the transport time will also be estimated with collapsed and scaled impedance spectroscopy and conductance data.

1. Zhou, R, Wang, P and Chang, H-C, â€œBacteria capture, concentration and detection by AC dielectrophoresis and self-assembly of dispersed single wall CNTs, *Electrophoresis*, **27**, 1375 (2006).
2. Basuray, S, Senapati, S, Aijian, A, Mahon, A R and Chang, H-C, â€œShear and AC field enhanced CNT impedance assay for rapid, sensitive and mismatch-discriminating DNA hybridization, *ACS Nano*, **3**, 1823 (2009).

472808 Rapid Isoelectric Focusing of Proteins in 1 Mm Long Microchannels

Abstract:

Isoelectric focusing (IEF) is used to separate proteins in millimeter long, agarose-filled microchannels. The agarose suppresses convection due to hydrostatic and electroosmotic flows that might otherwise occur in very short channels while loading or running the samples. Our experiments were performed in PMMA microchannels with lengths ranging from 1-10 mm at a constant field strength of 50 V/cm and these separations were recorded using whole channel imaging with fluorescent proteins. Narrow focused protein bands were usually observed in less than 1 min, but resolution slowly improved if several extra minutes were allowed to pass.

471712 Reversible Assembly of Colloidal Particles Using Low Frequency Pulsed DC Electric Fields for Electrophoretic Displays

Abstract:

We present the reversible assembly of colloidal nanoparticles in organic solvents using low voltage and low frequency pulsed DC electric fields. Commercially available and custom nanoparticles (diameter ~ 50-200 nm) were suspended in organic solvents at concentrations of 0.25-50 wt% and sandwiched between indium tin oxide (ITO) electrodes of spacing 25 Î¼m-2mm. Using square waves of frequency 0.25-5 Hz, voltages 0.5-5V, and DC offset 0.25-2.5V, colloidal suspensions were actuated for switchable contrast. Upon switching off the electric field, the assembled nanoparticles are able to resuspend within seconds. Using patterned electrodes, lateral electromigration of the particles was demonstrated. Depletion of the particles to the electrode regions resulted in transparency in the non-patterned regions. The distinct contrast generated can be exploited for tunable display applications. We will also present a particle-based numerical model of the deposition and resuspension processes.

This work was performed under the auspices of the U.S. Department of Energy by Lawrence Livermore National Laboratory under Contract DE-AC52-07NA27344.

472513 Production of Nanodrops Using Interfacial Electrokinetic Polarization at a Flow-Focused Microfluidic Constriction

Abstract:

Nanoemulsions are metastable dispersions of nanoscale droplets of one liquid suspended in a different immiscible liquid. Because they offer enhanced material properties and versatility, there is growing interest in their use in a wide variety of applications including the food and beverage, pharmaceutical, energy and cosmetic industries. Conventional high pressure liquid homogenizers and microfluidizers can produce nanoemulsions using high pressure to provide large mechanical stresses to break large droplets down into smaller sized nanoscale dispersions. One major problem with these methods is that they are expensive and require extremely large pressures – up to 20,000 psi – to create the required fluid shear stresses to form nanosized droplets. Alternatively, low pressure approaches such as self-emulsification require high surfactant concentrations, which can be undesirable or toxic in pharmaceutical or food-based applications. In this work we present a new electrokinetic method for producing nanoscale droplets. Unlike previous electro-spray emulsification techniques that promote droplet break-up using electrical shear stresses tangential to the flow direction, our approach combines classic microfluidic flow-focusing with downstream orthogonalelectrical stresses to produce droplets. In this way, the hydrodynamic and electrical influences are effectively decoupled and can be manipulated independently to optimize droplet emulsification. We show that nanosized droplets can beproduced by forcing a thin hydrodynamic jet of discontinuous fluid through a thin flow constriction. Within the constriction, we use locally patterned 3D carbon black electrodes to deliver a direct current (DC) electric field perpendicular to the direction of flow. The field induces an interfacial electrical stress at the discontinuous-continuous fluid phase interface, where the combined tangential-hydrodynamic and orthogonal-electrical stress forces the thin jet to break into small nanoscale drops.Using this method, we present data illustrating the ability to break thin liquid jets into sub-micron droplets and highlight important parameters such as elongational shear stress, electric field strength, surfactant concentration, and device geometry, that influence droplet diameter and show how our approach provides a novel low-cost scalable method for producing nanoscale emulsions at low pressure.

455140 Placement and Separation of Colloids By Liquid Crystal Enabled Electrokinetics Controlled By Patterned Substrates

Abstract:

Colloids self-assemble into various organized superstructures determined by particle interactions. There is a tremendous progress in both the scientific understanding and applications of self-assemblies of single-type identical particles. Forming superstructures from two or more different types of particles and separating different species under fixed experimental conditions represent two major challenges of the current state-of-the art. Here we propose a versatile approach to direct heterogeneous self-assembly and separation of colloids using liquid crystals with spatially varying molecular orientation pre-imposed by substrate photoalignment and electrokinetic flows in these systems. Colloidal spheres that differ in surface properties (anchoring of the local director) are driven by the gradients of molecular orientation into different locations where they self-assemble through long-range elastic forces. An applied AC electric field separates the spheres according to their surface properties, by steering them into opposite directions through the effect of liquid crystal-enabled electrokinetics. Elastic forces responsible for heterogeneous assembly and separation are measured by exploring overdamped dynamics of the colloids. Control of colloidal self-assembly and sorting through patterned molecular orientation opens new opportunities for design of reconfigurable materials, microfluidics, lab-on-a-chip applications, targeted cargo delivery, biological sensing and sorting. The work is supported by NSF grants DMR-1507637, DMR-1121288 and DMS-1434185.

470258 Optoelectric Trapping: Effect of Electrode Material and Thickness on Light-Induced Electrothermal Flow

Abstract:

Microfluidics is ushering a new era in drug discovery, tissue engineering, and point-of-care diagnostics. By conducting analysis in micro- to nanoscale environments, microfluidic biochips outperform some of the conventional benchtop technologies. However, these advantages come with the added complexity of particle manipulation at these small length scales; hence, since the advent of microfluidics, non-contact particle manipulation methods have been highly sought after. Rapid Electrokinetic Patterning (REP) has received considerable attention for its tunable trapping and translation of particles in a microfluidic chip. REP utilizes a laser-induced toroidal electrothermal flow and particle-electrode interactions for trapping particles on demand. In previously demonstrated applications of REP, a monochromatic infrared laser beam, integrated into an inverted microscope, was used to create the necessary temperature rise required for producing an electrothermal flow. However, usage of a high-intensity laser beam in a REP set up is expensive and time-consuming, restricting the adoption of the technique. The experimental setup will be significantly simplified if REP trapping can be achieved with an incoherent low-intensity light source. To accomplish this objective, we identify the optimum electrode layer so that REP can be performed with minimal optical-intensity. Using theoretical, computational, and experimental approaches, we demonstrate that conventionally used 700 nm Indium tin oxide (ITO) layer is an inefficient electrode choice as more than 92% of the irradiated illumination on the ITO electrodes undergoes transmission without absorption. At a given optical power, the magnitude of the electrothermal flow is controlled by both the electrode material and its thickness. After comparing electrode films of different materials and varying thickness, we find that a 25 nm thick titanium layer is the best electrode choice. This titanium electrode generates an electrothermal flow of the same magnitude as the ITO electrode while requiring only 14% of the light intensity used by ITO. In this presentation, we plan to discuss the importance of the electrode material choice and the associated applications.

DISPENSING SURFACTANT-CONTAINING WATER DROPLETS USING ELECTROWETTING

B. P. Chock,[a*] T. B. Jones,[b] and D. R. Harding[a]
[a]Laboratory for Laser Energetics, University of Rochester, Rochester, NY, USA 14623-1299
[b]University of Rochester, Rochester, NY, USA

ABSTRACT

Many droplet-based microfluidic schemes use the electrowetting-on-dielectric (EWOD) mechanism to dispense and transport droplets of conductive liquids. In applications where surfactant must be added to water, there are serious impediments to the droplet-dispensing process. The problem is a persistent membrane of fluid that forms between the reservoir and the droplet, spanning the top and bottom electrode surfaces. As the surfactant concentration approaches the critical micelle concentration, rupturing this membrane becomes increasingly difficult. This problem is resolved by embedding a thin resistive-heater element into the lab-on-chip device to assist the EWOD forces in creating droplets. When a dc current is applied to the heater strip, the membrane destabilizes and releases the droplet within ~0.5 s.

INTRODUCTION

Electric-field–based microfluidics has seen a rapid growth of interest, especially in the biomedical field. There are also recent efforts, including through this paper, to apply this type of microfluidics towards the fabrication of fuel capsules for fusion energy. A main reason for this is the ability to dispense, transport, and merge microliter-sized droplets to form emulsions. Chemicals in the outer phase of the emulsion can then be reacted to form the plastic capsules used to hold the deuterium-tritium fuel. The water-based fluids and oils used for the emulsion can be controlled using electrowetting on dielectric (EWOD) and dielectrophoresis, respectively.[1]

For the biomedical field, many aqueous solutions contain salt ions, protein, and biological cells.[2] The addition of a surfactant has arisen to prevent surface adsorption of the cells and proteins to solid surfaces. For the fusion-energy community, a surfactant is necessary to stabilize the layers in oil–water double emulsions, where key factors include the type and concentration of the surfactant.[3]

One serious problem with the microfluidics technology is that surfactant-water droplets are difficult to dispense in air; the problem stems from surfactant molecules inhibiting the separation of a droplet from a larger reservoir of fluid. During the dispensing process, a thin persistent membrane forms between the droplet being dispensed and the reservoir, the stability of which is related to the surfactant concentration relative to the critical micelle concentration (CMC). Attempts to rupture the membrane have included adding acid to the water, locally diluting a separating droplet with pure water, or using stronger electric fields.[1,4]

We report here on a new scheme to produce surfactant-water droplets in air using localized resistive heating to destabilize the membrane by thermocapillary effects.[5] This method facilitates rapid droplet production, lower operating voltages, and better longevity of the device. In this paper, the efficacy of dispensing droplets with resistive heating is examined for surfactant concentrations up to 5× the CMC limit.

PRINCIPLE OF OPERATION

EWOD

Electrowetting on dielectric (EWOD) is an electromechanical mechanism used to move and manipulate conductive liquid droplets such as water at a low frequency of 100 Hz and relatively low voltages (<100 V_{rms}). An electrical force, resulting from the electric field acting on the surface charge near the dielectric/liquid/air contact line, causes the contact line to spread across the activated electrode.[6] The lab-on-chip device acts as a capacitor and the change in capacitance (C) with respect to the horizontal distance (x) along an energized electrode provides a way to model the net electrical force on a droplet (F):[7]

$$F_{\text{electric}} = \frac{V_{\text{rms}}^2}{2}\left(\frac{dC}{dx}\right). \tag{1}$$

Materials and Experimental Setup

Parallel-plate electrode designs are wet-etched into an aluminized glass substrate. The substrate is then spin-coated with a 0.3-μm spin-on-glass suspension layer (Futurrex) followed by a 0.9-μm Teflon® layer (6 wt%) as received from DuPont. An indium-tin-oxide (ITO)–coated glass slide, spin-coated with a 0.9-μm Teflon® layer, serves as the ground electrode. Two sets of substrate geometries are used: (1) 2 × 2-mm electrodes with 90-μm substrate spacing between parallel plate; and (2) 6 × 6-mm electrodes with 270-μm substrate spacing. Spacers between the base and top electrodes provide the desired spacing. For the experiments, 75 V_{rms} at 10 kHz (high-frequency EWOD) is used for general operations while 10 to 30 V_{dc} is applied to the heater.

Silwet L-77 (a non-ionic surfactant suspended in solvent), used as received from the supplier (Momentive), is diluted with de-ionized (DI) water (18.2 MΩ-cm). The range of surfactant concentrations is 0.005, 0.01, 0.025, 0.05 wt%, where the CMC is ~0.01 wt%. These concentrations translate to 0.5×, 1×, 2.5×, and 5× the CMC limit, respectively.

EXPERIMENTAL

Dispensing Droplets with Resistive Heating

The problems associated with dispensing water droplets with higher surfactant concentrations can be substantially reduced when a resistive heater is embedded into the electrode design (Fig. 1). First, using EWOD forces, fluid is pinned at the droplet electrode, while fluid is retracted from the cutting electrode to the reservoir electrode. When the enclosing menisci at the cutting electrode meet, they form a membrane. Applying a dc current to the heater electrode, the generated heat destabilizes the membrane.

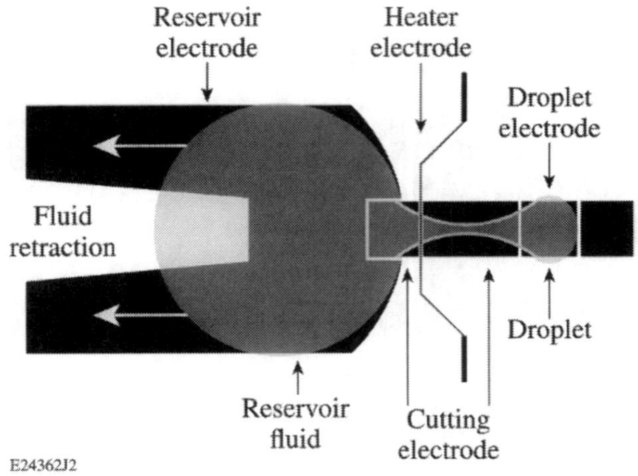

Fig. 1. Schematic (top view) of the lab-on-chip electrode design. The fluid on the powered droplet-formation electrode is attached to the reservoir fluid by a thin membrane. Black regions correspond to the reservoir, cutting, heater, and droplet electrodes. Blue regions correspond to the fluid.

This concept is demonstrated in Fig. 2 for a 1× CMC solution droplet being dispensed at 75 V_{rms} and 10 kHz. At this concentration, the membrane was stable (when no heat was applied). In Figs. 2(a)–2(d), the width of the electrodes was 2 mm and the substrate spacing between plates was 90 μm. At 8 s prior to the t = 0-s frame, the fluid at the cutting electrode retracts to the reservoir electrode. At t = 0 s, current was applied to the heater (the effect of heating is noted by red arrows). Within 0.33 s, the membrane ruptured and the droplet formed.

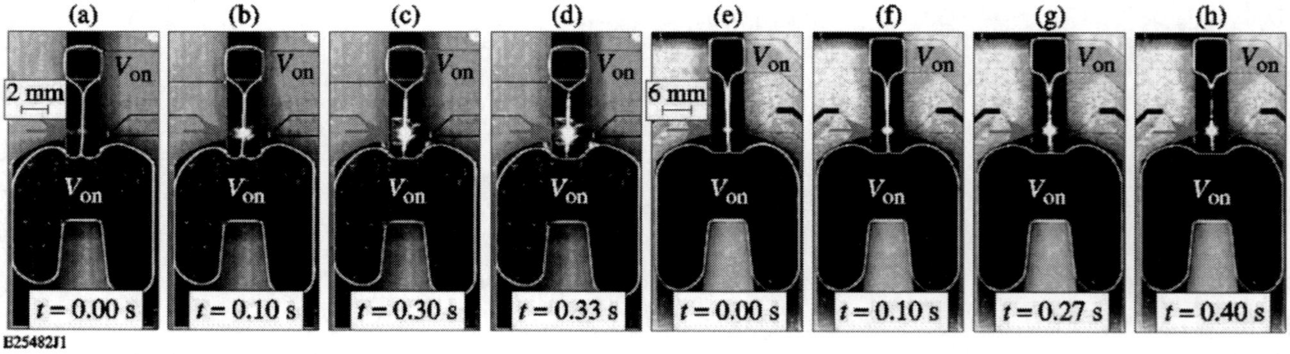

Fig. 2. Photograph sequence (top view) of the fluid membrane for a 1× CMC solution being locally heated by a resistive strip (30-V_{dc} application). [(a)–(d)] The electrode width is 2 mm and the substrate spacing is 90 μm. The membrane ruptures after ~0.33 s. [(e)–(h)] The electrode width is 6 mm and the substrate spacing is 270 μm. The membrane ruptures after ~0.27 s.

A 30-V_{dc} potential across the V-shaped aluminum trace provided the heat (~8.11 mW with R_{meas} ~ 306 Ω). At the site of the membrane, the rate of heat generation can be approximated using the following equations:

$$\text{heat generation} = I^2 R_N (L_{NM}/L_N)\text{, where } L_{NM} \approx 0.184 \text{ mm,} \tag{2}$$

$$R_N = R_{meas} \frac{L_N}{W_N} \left(\frac{L_{W_1}}{W_{W_1}} + \frac{L_{W_2}}{W_{W_2}} + \cdots + \frac{L_N}{W_N} \right)^{-1}, \tag{3}$$

$$I = V_{dc}/R_{meas}, \tag{4}$$

where I is the current; R_N is the resistance of the straight, thin section of the electrode; R_{meas} is the resistance of the entire heater electrode; L_{NM} is the length beneath the membrane; L_N is the length of the straight, thin section of the heater electrode; $L_{W_1}, L_{W_2}, \cdots$ are the lengths of the neighboring sections; and W_N and $W_{w_1}, W_{w_2} \cdots$ are the electrode widths of the thin, straight and neighboring sections, respectively.

For Figs. 2(e)–2(h), the electrode width was 6 mm and the substrate spacing was 270 μm; 30 V_{dc} was applied across the heating trace, and the heat generated at the site of the membrane was ~39.4 mW (with R_{meas} ~ 113 Ω). The membrane ruptured within 0.4 s after heat was first applied.

The destabilization process for the membrane was investigated by focusing the camera lens on the heated section of the membrane under reduced lighting conditions. In Fig. 3 (operation at 75 V_{rms}, 10 kHz), 30 V_{dc} was applied to the heating trace for surfactant-water solutions at 1× and 5× the CMC limit. The electrode width was 6 mm and the substrate spacing was 270 μm. In the first frames of each run (t = 0 s), a dc voltage was applied; the location of the heat trace could be pinpointed by the circular dot. For the 1× CMC concentration [Fig. 3(b) (~0.13 s of heating)], there were indications of wave fluctuations along the entire length of the membrane. The fluctuations continued to grow in Figs. 3(c) and 3(d), and the membrane ruptured after ~0.23 s to yield a pearl string of satellite droplets. For the membrane of a 5× CMC solution, unstable surface waves appeared along the membrane [Figs. 3(h)–3(j)], and the membrane ruptured after ~0.27 s.

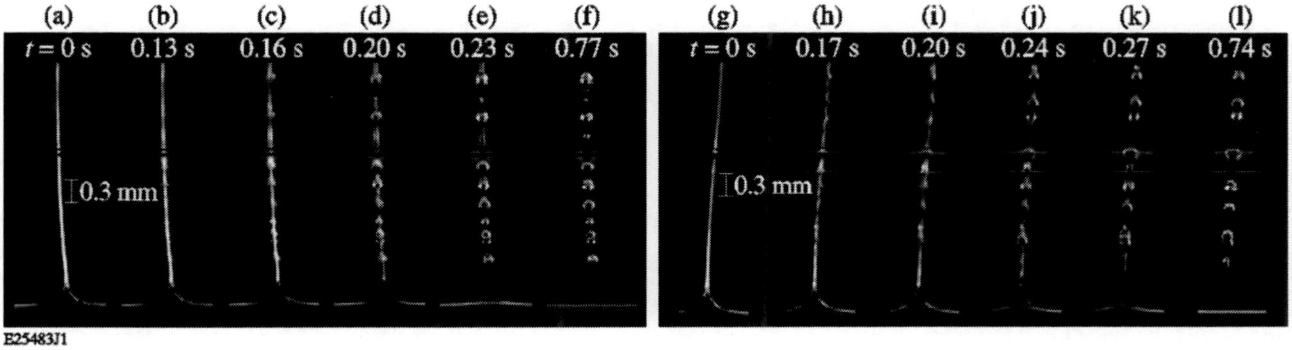

Figure 3: Photograph sequences (top view) of the breakup of a surfactant-water membrane being heated from below using 30 V_{dc}. The electrode width was 6 mm, and the substrate spacing was 270 μm. [(a)–(f)] 1× CMC solution; [(g)–(l)] 5× CMC solution.

One can observe from the 1× and 5× CMC runs that the entire membrane appears to destabilize and rupture simultaneously. The breakup pattern for the membrane seems to differ from that suggested by Yeo,[5] who predicted that the membrane would first rupture at the site of heating (only a single substrate was used in his simulations). As discussed by Cho,[8] the top substrate, which is the ground electrode, may be contributing to the stability of the membrane.

DISCUSSION

Pressure Analysis of a Liquid Stream

The breakup of a liquid stream in air shares a common underlying mechanism with the breakup of a liquid membrane during the droplet-dispensing process. Consider the Rayleigh–Plateau instability of a cylindrical liquid column as shown in Fig. 4.[9] Regions where the column is narrower (P_1) develop a higher static pressure than regions where the column is wider (P_2) because of the Laplace pressure. Consequently, fluid is forced from narrower regions to wider regions and droplets are pinched off. If P_{air} is the ambient pressure and Υ_{LV} is the surface tension, then[8,10]

$$P_1 = P_{air} + \Upsilon_{LV}\left(\frac{-1}{R_1} + \frac{1}{r_1}\right), \tag{5}$$

$$P_2 = P_{air} + \Upsilon_{LV}\left(\frac{1}{R_2} + \frac{1}{r_2}\right). \tag{6}$$

Many EWOD-related papers end their discussion for the breakup of the membrane at this point, but simple pressure differences are inadequate to explain membrane persistence when surfactant is added.[1,8] For thin membranes, surfactant molecules contribute an important disjoining pressure component.

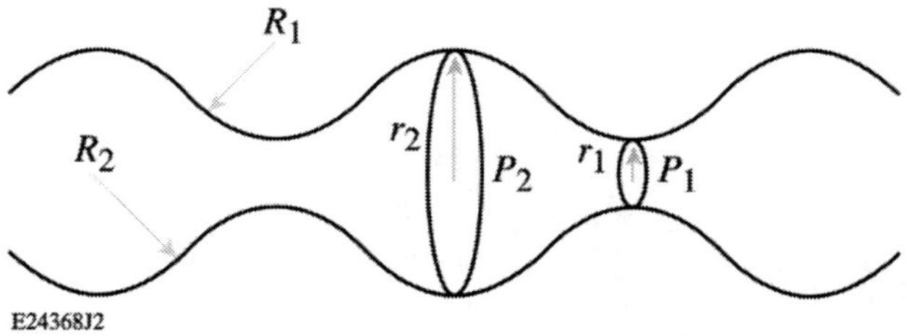

Fig. 4. Schematic of a liquid stream in air showing regions of higher (P_1) and lower pressure (P_2) resulting from the fluctuating thickness and the corresponding Laplace pressure (R_i, r_i = radii of curvature).

Disjoining Pressure and Apparent Film Pressure

As a liquid membrane thins, it becomes a film; in such a case, there is more to consider than the Laplace pressure. At thicknesses of ~100 nm or less, molecular interactions are affected by the proximity of the two liquid/air interfaces. The result is a disjoining pressure (Π) contribution that changes the pressure within the thin film [referred to as the "apparent pressure" (P_f')] relative to the static pressure in the bulk liquid:[11]

$$\text{apparent pressure} = P_{\text{liquid}} + \Pi . \tag{7}$$

The disjoining pressure [$\Pi(h)$], a function of h (the membrane thickness), includes Van der Waals (Π_{VDW}), electrostatic (Π_{elec}), structural ($\Pi_{structural}$), and short-range (Π_{short}) intermolecular forces acting between the surfaces of the liquid film:[11,12]

$$\Pi(h) = \Pi_{VDW} + \Pi_{elec} + \Pi_{structural} + \Pi_{short} . \tag{8}$$

The Van der Waals pressure component is negative, the electrostatic term is generally positive, the structural term is normally positive, and the short-range term varies in sign. At the film/air interface, surfactant molecules greatly increase the magnitude of the electrostatic term, while micelles in solution increase the structural term. The effect of the short-range term, which accounts for the interaction force between the two enclosing interfaces at a thickness of ~5 nm, is less understood.[11,13]

The ability to rupture the film is related to the pressure difference (ΔP) between the air (P_{air}) and the liquid membrane (P_f'). If $\Delta P > 0$, the membrane ruptures:

$$\Delta P = P_{\text{air}} - P_f' . \tag{9}$$

Wave Instability Growth for Liquid Films (No Heating)

An alternative model for the breakup of a draining film[14] considers wave fluctuations along the membrane as fluid is evacuated similar to the schematic from Fig. 4. The surface waves, created by mechanical perturbations or thermal fluctuations, are unstable for wavelengths greater than a critical value:

$$\lambda_{\text{crit}} = 2\pi\sqrt{\frac{\pi \Upsilon h^4}{A}} , \tag{10}$$

where Υ is the surface tension of the fluid, h is the average thickness of the film, and A is the Hamaker–Van der Waals constant.[14] Fluctuations with wavelengths below this critical value will diminish, but those above it are unstable. As the film thins, the critical wavelength to achieve instability decreases and it is easier to rupture the film.

Thermocapillary Effects When a Liquid Film Is Uniformly Heated

If a thin film resides on a uniformly heated surface, surface undulations will exist owing to temperature gradients at the liquid/air interface.[5,15] Regions of the film surface closer to the heated substrate have a higher temperature than those farther away. The surface tension of the film will be lower in the hotter regions and greater in the cooler regions, and fluid will flow from the hot to cold regions.[16] As a result, thermocapillary instabilities may develop.

The thermal Biot (Bi) number at the air/liquid surface compares the ratio of the resistance to heat transfer externally (convection) versus internally (diffusion):

$$\text{Bi} = \frac{\alpha_h H}{k} = \frac{H/k}{1/\alpha_h}. \tag{11}$$

Here α_h is the heat transfer coefficient at the air/liquid interface, H is the initial film thickness, and k is the thermal conductivity of the liquid film. Therefore, there is a second critical wavelength (λ_c) that depends on the Biot number, where the surface waves along the film become unstable:[5]

$$\lambda_c = 2\pi\left[\frac{2C(1+\text{Bi})^2}{3\text{Bi}}\right]^{1/2}, \tag{12}$$

i.e., C is a capillary parameter and larger values denote increased capillary effects.

Thermocapillary Effects When a Liquid Film Is Nonuniformly Heated

If, instead, a thin liquid film rests on a substrate that is nonuniformly heated, as in the situation of interest here, fluid on top of the heated region will flow to the nonheated regions. The heated section becomes thinner than the colder section and a wave structure is created.[5,17] As the Biot number increases from 0.01 to 1, the amplitude of these waves increases (and the displacement of fluid increases). Further increases in the Biot number, however, result in the wave amplitude steadily decreasing.

The Biot Number and Experiments with Nonuniform Heating

From Figs. 2(a)–2(d) and Figs. 2(e)–2(h), there is a threefold difference in substrate spacing between plates (90 μm versus 270 μm). According to Eq. (11), for the case in which the electrode width is 2 mm, the Biot number is expected to be threefold larger than when the electrode width is 6 mm (assuming the heat-transfer coefficient and the thermal conductivity are constant). Therefore, for Fig. 2 it makes sense that the fluid membrane of a 1× CMC solution (for the 90-μm spacing) requires less heating (~8 mW versus ~39 mW) to achieve a similar breakup time compared to the 270-μm case.

Proposed Mechanism for Rupturing the Membrane with Nonuniform Heating

Figure 5 depicts the breakup process for a draining liquid membrane with nonuniform heating. Before heat is applied [Fig. 5(a)], the draining membrane appears flat but has small surface fluctuations, owing to fluid motion. Fluid is transferred along the length of the surfactant-stabilized membrane. When heat is applied to the membrane [Fig. 5(b)], the heated liquid flows toward the neighboring sections of the membrane, owing to the gradient in temperature. This creates a thin section at the site of heating, while the adjacent sections become thicker. Then, from the newly thickened regions, fluid flows to the neighboring (colder) sections of the membrane, which in turn become thicker [Figs. 5(c) and 5(d)]. The result is that the membrane becomes thinner as a whole [Fig. 5(e)]. The drainage process described in Figs. 5(a)–5(e) should be faster when more heat is applied to the membrane (larger Biot number).

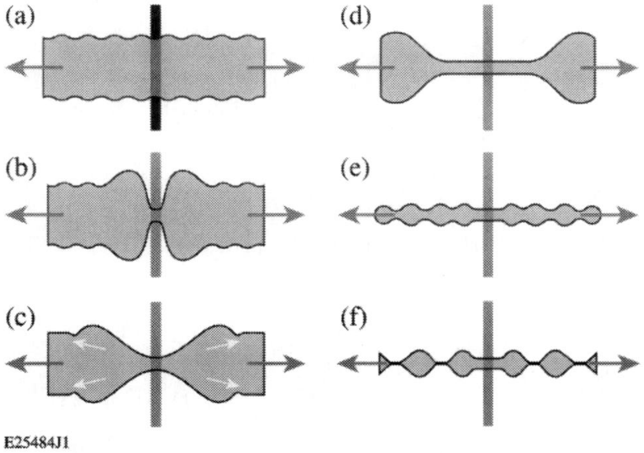

Figure 5: Schematic (top view) of a draining thin membrane being heated from below by a resistive heater. (a) No voltage is applied to the resistive strip, and the membrane has small wave fluctuations, owing to fluid being transferred through the film to the powered electrodes in either direction. (b) A dc potential has been applied across the resistive heating strip, and the fluid on top of the strip begins to heat up. The fluid flows from the heated regions to the adjacent regions. (c) The movement of fluid from the heated region warms the neighboring sections of the membrane. Fluid then flows from those sections of the membrane to sections farther away from the site of heating. (d) The continued flow of fluid away from the site of heating causes the membrane to become thinner. (e) The membrane becomes thinner and unstable waves begin to form along the membrane. (f) The membrane continues to drain and will soon rupture.

For Fig. 5(e), consider Eqs. (10) and (12): It is easier for the surface waves along the membrane to become unstable if the membrane is thinner. More-rapid heating may also make it easier for the surface waves to destabilize (by decreasing the critical wavelength for instability). When the instability along the membrane has adequately grown [Fig. 5(f)], the membrane will rupture, owing to Rayleigh–Plateau instability, which accounts for Van der Waals interactions.

Breakup Time of the Membrane with Heat Application

Figure 6 plots the heat application times for membranes at 1×–5× the CMC limit for an electrode width of 2 mm and substrate spacing between plates of 90 μm. The time needed to rupture the membrane significantly decreases when the heater voltage is increased from 10 V to 30 V (the heat applied increases from 0.90 mW to 8.11 mW). For any given applied dc voltage, the time scales to destabilize the membranes of different surfactant concentrations are similar.

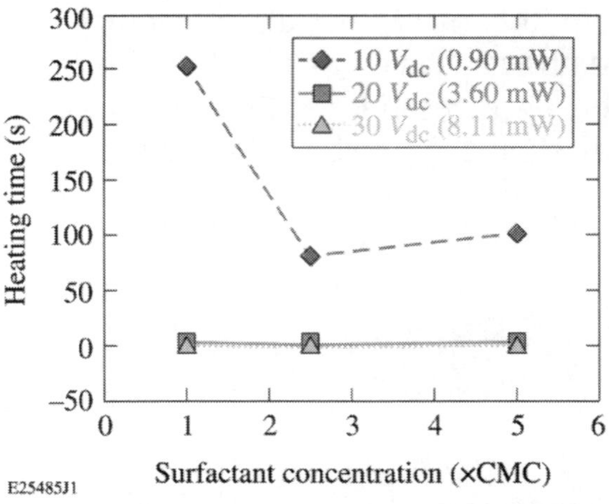

Fig 6: Estimated time the membrane is heated by the resistive heater for different dc voltages. The electrode width is 2 mm and the substrate spacing is 90 μm.

In Fig. 7 the heating times for membranes at 0.5× to 5× the CMC limit are plotted when the electrode width is increased to 6 mm and the substrate spacing is 270 μm. A voltage of 20 V_{dc} or 30 V_{dc} is applied to the heater trace (17.5 mW and 39.4 mW of heating, respectively) and the heating time is always less in the 30-V_{dc} case. In addition, the necessary application time for a 0.5× CMC solution is substantially smaller than for the 1×–5× CMC cases, and the shorter time may be related to the amount of surfactant at the air/liquid interface. The heating times for the 1×–5× CMC concentrations are similar.

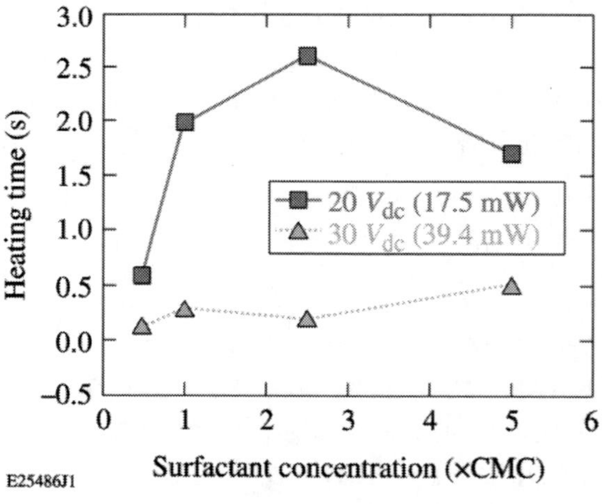

Fig 7: Estimated time for which the membrane is heated by the resistive heater for different dc voltages. The electrode width is 6 mm and the substrate spacing is 270 μm.

Droplet Volumes and dc Voltage

The volumes of the dispensed droplets are affected by the magnitude of the dc voltage applied to the heater. From Fig. 8 (2-mm electrode width, 90-μm spacing) the droplet volumes (at 1×–5× the CMC limit) range from 0.36 μL to 0.68 μL. When 10 V_{dc} is applied for the 1× and 5× CMC concentrations, the droplet volumes are small. The longer heating time needed to rupture the membrane may be a factor since fluid will continue to drain from and through the membrane. At higher dc voltages, the volumes produced are similar for all of the surfactant solutions (1×–5× the CMC limit). This trend may be related to the shorter heating time (Fig. 6) in each case.

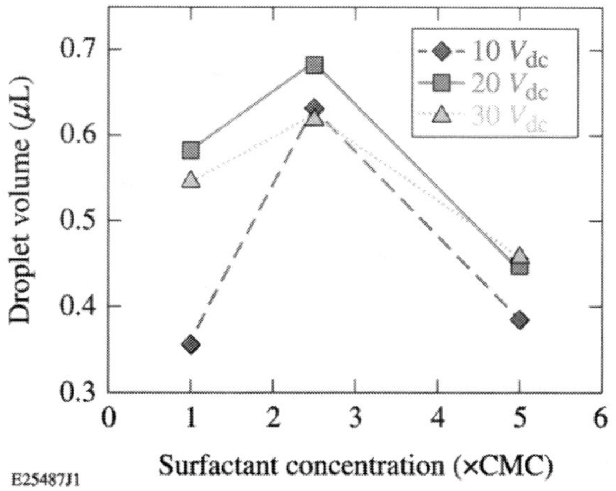

Fig. 8: Droplet volumes formed with a range of dc voltages (10 V_{dc} to 30 V_{dc}) applied to the resistive heater. The electrode width is 2 mm and the substrate spacing is 90 μm.

From Fig. 9 (6-mm electrode width, 270-μm spacing) the droplet volumes (at 0.5×–5× the CMC limit) range from 10.6 μL to 13.4 μL when either 20 V_{dc} or 30 V_{dc} is applied. When 20 V_{dc} is applied (versus 30 V_{dc}) the dispensed droplet volumes generally decrease. It is likely that the lower heating rate (17.5 mW versus 39.4 mW) allows for additional fluid to drain out of the membrane.

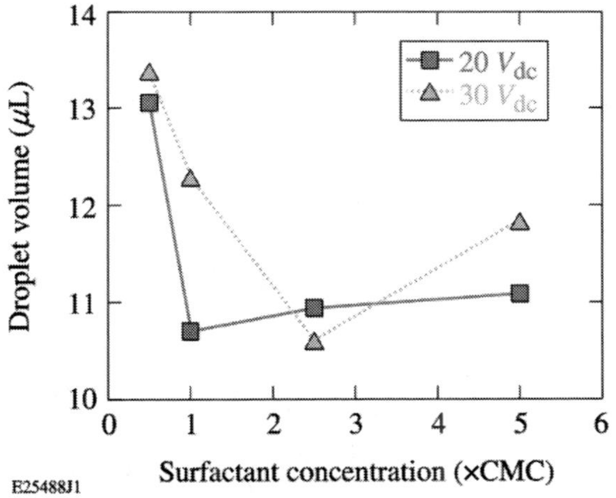

Fig. 9: Droplet volumes formed with a range of dc voltages (20 V_{dc} to 30 V_{dc}) applied to the resistive heater. The electrode width is 6 mm and the substrate spacing is 270 μm.

CONCLUSION

Surfactant-water droplets (~0.36 μL to 13.4 μL) are produced for surfactant concentrations from 0.5× to 5× the CMC limit with 10 V_{dc} to 30 V_{dc} applied to the heater electrode and 75 V_{rms} applied for general operations. The lower ac voltage reduces the risk of dielectric breakdown; meanwhile the resistive heating (10 V_{dc} to 30 V_{dc}) helps to destabilize the membrane by decreasing its thickness. By varying the dc voltage from 10 V to 30 V, it has been demonstrated that the shortest time to destabilize the membrane is at 30 V. The droplet volumes that are dispensed are generally similar when the applied dc voltage is 20 V or 30 V.

For the breakup of a nonuniformly heated membrane (Fig. 3), it has been revealed that the whole membrane destabilizes before it ruptures. This is a departure from thermocapillary theory predictions for the breakup of a liquid film. It is likely that the two substrates that bound the fluid help to stabilize the membrane during the heating process. Only after the membrane has thinned sufficiently are the stabilizing components of the disjoining pressure and surface tension overcome by Van der Waals interactions, making it possible for the membrane to rupture.

ACKNOWLEDGMENT

This work was supported by the Frank J. Horton Graduate Fellowship Program. This material is based upon work supported by the Department of Energy National Nuclear Security Administration under Award Number DE-NA0001944, the University of Rochester, and the New York State Energy Research and Development Authority. The support of DOE does not constitute an endorsement by DOE of the views expressed in this article.

REFERENCES

1. Wang, W. (2012), "On-Chip Double Emulsion Droplet Assembly Using Electrowetting-on-Dielectric and Dielectrophoresis," Ph.D. thesis, University of Rochester.

2. Shah, G. J., J. L. Veale, Y. Korin, E. F. Reed, H. A. Gritsch, and C.-J. Kim (2010), "Specific Binding and Magnetic Concentration of CD8+ T-Lymphocytes on Electrowetting-on-Dielectric Platform," Biomicrofluidics, **4**, 044106; Miller, E. M., A. H. C. Ng, U. Uddayasankar, and A. R. Wheeler (2010), "A Digital Microfluidic Approach to Heterogeneous Immunoassays," Anal. Bioanal. Chem., **399**, pp. 337–345.

3. Brooks, B. W. and H. N. Richmond (1991), "Dynamics of Liquid—Liquid Phase Inversion Using Non-Ionic Surfactants," Colloids and Surfaces, **58**, pp. 131–148.

4. Chock, B. P., T. B. Jones, and D. R. Harding (2016), "Effect of a Surfactant on the Electric-Field Assembly of Oil-Water Emulsions for Making Foam Targets," Fusion Sci. Technol. **70**, pp. 206–218.

5. Yeo, L. Y., R. V. Craster, and O. K. Matar (2003), "Marangoni Instability of a Thin Film Resting on a Locally Heated Horizontal Wall," Phys. Rev. E, **67**, 056315.

6. Jones, T. B. (2009), "The Electromechanics of Liquids," presented at Microfluidics: Electrokinetic and Interfacial Phenomena, University of Minnesota, Minneapolis, MN, 7–11 December.

7. Jones, T. B. (2005), "An Electromechanical Interpretation of Electrowetting," J. Micromech. Microeng., **15**, pp. 1184–1187.

8. Cho, S. K., H. Moon, and C.-J. Kim (2003), "Creating, Transporting, Cutting, and Merging Liquid Droplets by Electrowetting-Based Actuation for Digital Microfluidic Circuits," J. Microelectromech. Syst., **12**, pp. 70–80.

9. de Gennes, P.-G., F. Brochard-Wyart, and D. Quere (2004), "Capillarity and Wetting Phenomena: Drops, Bubbles, Pearls, Waves," Springer, New York.

10. Gong, J. and C.-J. Kim (2008), "All-Electronic Droplet Generation On-Chip with Real-Time Feedback Control for EWOD Digital Microfluidics," Lab Chip **8**, pp. 898–906.

11. Tadros, T. F. (2005), "Applied Surfactants: Principles and Applications," Wiley-VCH, Weinheim, Germany.

12. Stubenrauch, C. and R. von Klitzing (2003), "Disjoining Pressure in Thin Liquid Foam and Emulsion Films—New Concepts and Perspectives," J. Phys., Condens. Matter, 2003, **15**, pp. R1197-R1232.

13. Ivanova, D. S. and J. K. Angarska (2013), "Drainage and Critical Thickness of Foam Films Stabilized by *n*-Dodecyl-β-D-Maltoside," Colloid Surf. A, Physiochem. Eng. Aspect, **438,** pp. 93–103.

14. Vrij, A. and J. Th. G. Overbeek (1968), "Rupture of Thin Liquid Films Due to Spontaneous Fluctuations in Thickness," J. Am. Chem. Soc., **90** pp. 3074–3078.

15. Chinnov, E. A. and O. A. Kabov (2007), "Marangoni Effect on Wave Structure in Liquid Films," Microgravity Sci. Technol., **19**, pp. 18–22.

16. Brochard, F. (1989), "Motions of Droplets on Solid Surfaces Induced by Chemical or Thermalgradients," Langmuir, **5**, pp. 432–438.

17. Kalpathy, S. K., L. F. Francis, and S. Kumar (2013), "Thermally Induced Delay and Reversal of Liquid Film Dewetting on Chemically Patterned Surfaces," J. Colloid Interface Sci., **408**, 212–219.

464969 Electric Field-Driven Structuring in Suspensions

Abstract:

Emerging routes based on electro-micro-fluidic systems enable the rapid fabrication of colloidal materials without the need for expensive tooling and offer exciting opportunities for materials assembly. Although each one of such technologies is independently unique, all of them share the common feature of utilizing dielectrophoresis and structural transformations, governed by the dipolar interactions, which vary from attraction to repulsion, depending on the relative orientation of the dipoles. The ability to manipulate particles is based on the particle-fluid polarization contrast that characterizes a difference in polarizability between particles and a host fluid and depends on the mismatch of their dielectric constants and conductivities and the field frequency. Advantages include the use of an electric field that requires no moving parts, the employment of the polarization force acting on a particle that is insensitive to its charge which is difficult to control, the provision of two independent control variables (field strength and frequency) and the ease of adaptability to electronics, so that it can be incorporated more favorably into micro-and nano-systems. The widespread use of electrically driven colloidal processes emphasizes a critical need for improving fundamental understanding of the role of spatially and time varying fields in directing and controlling non-equilibrium phenomena in a suspension of interacting colloidal particles. The challenge is due to *kinetic limitations* because the energy of the interparticle dipolar interactions at typical field strengths is tens to hundreds of times stronger than the Boltzmann energy. The colloid can therefore be kinetically trapped into metastable configurations for a long time due to the lower mobility of multi-particle structures formed in an applied field compared to that of individual particles. Although numerous experimental studies have been performed involving "electrically tunable" dipole-dipole interactions, the rich variety of the patterns that have been described could also have been influenced by gravity effects, such as particle sedimentation, convection and jamming, which often compete with dipolar forces during the slowly evolving structure formation.

Flight tests conducted in NASA Zero-Gravity airplane demonstrate that the presence of even *weak gravity* forces in a suspension of *not density matched* particles *significantly changes* dielectrophoretic processes and structure formation. A requirement for *precise matching* of densities between particles and a host fluid in order to avoid undesirable gravity effects severely limits the possibilities for varying *a particle-fluid polarization contrast* in terrestrial experiments.

Microgravity offers an opportunity to probe electrically driven phenomena *over a wide range* of the mismatch of dielectric constants and conductivities between particles and a host fluid by conducting experiments on *not density matched* suspensions in long timescales without *masking effects of the gravity.* We will report experimental data on field driven structure formation in neutrally-buoyant suspensions and present our approach to the development of experiments in the Advanced Colloids Experiment facility of the International Space Station (ISS). The aim of the ISS experiments is to observe electrically driven phenomena in model suspensions of not density matched particles is to understand basic mechanisms of structure formation in polarized suspensions and to suggest novel routes for creating functional materials.

The phenomena selected for ISS tests are logical candidates to take advantage of microgravity because terrestrial observations failed to provide a comprehensive physical understanding of the underlying physics.
The work is supported by NASA's Physical Science Research Program, NNX13AQ53G.

466710 Electrohydrodynamics of a Viscous Drop with and without Inertia

Abstract:

In this talk we present our recent work on electrohydrodynamics of a viscous drop in an external electric field. In the first half we present work on spheroidal modeling of a surfactant-laden viscous drop under an electric field in the Stokes flow regime. In the second half we present numerical results from a recent study of the inertia effects on the electro-deformation of a viscous drop under a DC electric field using a novel second-order immersed interface method. The inertia effects are quantified by the Ohnesorge number Oh, and the electric field is characterized by an electric capillary number Ca_E. Below the critical Ca_E, small to moderate electric field strength gives rise to steady equilibrium drop shapes. We found that, at a fixed Ca_E, inertia effects induce larger deformation for an oblate drop than a prolate drop, consistent with previous results in the literature. Moreover, our simulations results indicate that inertia effects on the equilibrium drop deformation are dictated by the direction of normal electric stress on the drop interface: Larger drop deformation is found when the normal electric stress points outward, and smaller drop deformation is found otherwise. To our knowledge, such inertia effects on the equilibrium drop deformation has not been reported in the literature.

Above the critical Ca_E, no steady equilibrium drop deformation can be found, and often the drop breaks up into a number of daughter droplets. In particular our Navier-Stokes simulations show that, for the parameters we use, (1) daughter droplets are larger in the presence of inertia, (2) the drop deformation evolves more rapidly compared to creeping flow, and (3) complex distribution of electric stresses for drops with inertia effects.
Our results suggest that normal electric pressure may be a useful tool in predicting drop pinch-off in oblate deformations.

468505 Nano-Crater Formation on Electrodes during the Electrical Charging of Aqueous Drops

Abstract:

A water drop in an insulating fluid acquires charge when it contacts an electrode, but experimental measurements of the charge acquired by the drop have been plagued by irreproducibility and the exact mechanism of charge transfer has remained obscure. Previous work, dating back to Maxwell, has implicitly assumed that the electrode remains unaltered by the charging process. Here we demonstrate that, contrary to this assumption, water drops and other conducting objects create "nano-craters" on the electrode surface during the charging process. We used optical microscopy, SEM, and atomic force microscopy to characterize the electrode surfaces before and after water drops were electrically bounced on them. We show that each drop contact creates an approximately one micron wide and 50-nm deep crater on the electrode surface. Given enough time, the drop will form enough overlapping nano-craters to effectively "eat through" a sufficiently thin electrode. We provide visual evidence that dielectric breakdown occurs during each charge transfer event, and we interpret the nano-crater formation as a consequence of Joule heating and local melting of the metallic electrode. The observations reported here help explain the long standing discrepancies between theoretical predictions and experimental measurements of the charge transfer, and are of particular interest in the development of microfluidic devices that use electrodes to control the motion of aqueous drops.

462829 Addressing of Small Droplets in Systems of Two Aqueous Phases Mediated By Electric Field

Abstract:

Systems of two aqueous phases provide mild conditions in extraction of proteins and other biomolecules. These systems are typically formed either by two incompatible polymers dissolved in water (polyethylene glycol and dextran) or by a mixture of polymer molecules (e.g. polyethylene glycol), inorganic salt and water. Droplets of one phase can serve as containers of reactants in microfludic devices. One possibility how to address a droplet is to use an external electric field of proper orientation. We have found that electric charge is spontaneously formed at the interface of two aqueous phases. The interaction of the imposed electric field and electric double layer around a droplet leads to the electroosmotic slide and oriented movement of the droplet.

We have tested addressing of droplets in several systems of two aqueous phases in open and closed microfluidic devices. Polyethylene glycol (PEG), different inorganic salts and water were used to prepare aqueous two phase systems. Top and bottom phases were formed. The top phase is rich of PEG (PEG phase) and the bottom phase is rich of a dissolved inorganic salt (salty phase). We obtained the dependences of droplet velocity on the applied strength of electric field, droplet size, and droplet composition. It was found that droplets of the salty phase placed on PEG phase move much faster than droplets of PEG phase placed on the salty phase. The salty droplets moved always to the negatively charged electrode. The effective electrophoretic mobilities were determined for all studied two aqueous phase systems. Finally, the addressing of salty droplets at microchannel bifurcation and complex microfluidic structures will be presented. The way of efficient separation of electrode chambers and the main fluidic structure will be also shown.

474380 Pushing the Limits on High Resolution Electrophoretic DNA Separations on Microdevices with Short Effective Separation Lengths

Abstract:

The conventional method of forensic DNA analysis involves capillary electrophoresis (CE) of short tandem repeat (STR) fragments amplified from select locations in the genome. CE has been the modern workhorse for genetic analysis since the 1990â€™s, and the systems specifically manufactured for forensic STR profiling are expensive and reagent cost-intensive. In addition, separations are on the order of 30-40 minutes, require highly-skilled personnel and are not portable. There is a need for the development of an inexpensive, rapid, and portable device for DNA fragment separation. Here, we report progress towards the development of a unique, multi-layer, rotationally-driven microdevice to meet these needs.

We have demonstrated the fabrication of microdevices composed of inexpensive materials such as cyclic olefin copolymer (COC), polyester film (PE), pressure sensitive adhesive (PSA), and other proprietary materials not typically used in conventional methods. Through CAD design, we have

designed, developed and tested electrophoretic microfluidic devices with up to four separation domains on a rotationally-driven system the size of a compact disc. Where possible, a CO 2 laser was used ablate channels into the necessary layers to create intricate fluidic channels with features less than 100 µm, with plasma oxidization providing enhanced layer-to- layer bonding using only an office laminator. Separations of fragments in the 80-300 base range were achievable in less than 300 sec with a 2-base resolution in a 4 cm L eff channel. This has the potential to integrate with other microdevice domains such as extraction and PCR for rapid human identification/screening in the field.

474379 Nanopore Sensing - Beyond DNA Sequencing

Abstract:

Nanopores (NPs) are single molecule biosensors, which utilize electrokinetic focusing and transport to detect and characterize unlabeled biomolecules.[1] Controlling and tuning the capture rate and the translocation speed of biomolecules through the NP are crucial features to allow sensing of fine biomolecular features within the experimental bandwidth limitations. In the first part of my lecture I will discuss two physical methods to: (i) enhance the *capture rate of DNA* molecules into solid-state using salt gradients to enable sensitive sensing down to a few attomoles,[2] and (ii) slowing down the *translocation speed* of DNA and proteins through NPs using a novel optoelectronic effect that can be switched on or off in a fraction of a millisecond.[3]

In the second part of my lecture I will discuss biophysical applications utilizing NPs: (i) barcoding and distinguishing among two highly similar HIV genes using solid-state NPs by incorporation of sequence specific peptide-nucleic acids (PNA) probes[4]; (ii) discrimination among two, unlabeled, Ubiquitin (Ub) protein conformers, which only possess a slightly different 3D structure and having the same amino-acid sequence.[5] (iii) direct and label-free detection of individual transcription-factors bound to DNA.[6] The ability to analyze biomolecular complexes, fine features along DNA strands and small proteins in their native folded state at the single molecule level opens up broad opportunities in biomedical research and in biophysics.

References

1. Wanunu, M. & Meller, A. in *Laboratory Manual on Single Molecules* (eds. Selvin, P. R. & Ha, T.) Cold Spring Harbor Press (2008).

2. Wanunu, M., Morrison, W., Rabin, Y., Grosberg, A. Y. & Meller, A. Electrostatic focusing of unlabeled DNA into nanoscale pores using a salt gradient. *Nat Nanotechnol* 5, 160–165 (2010).

3. Di Fiori, N. *et al.* Optoelectronic control of surface charge and translocation dynamics in solid-state nanopores. *Nat Nanotechnol* 8, 946–951 (2013).

4. Singer, A., Rapireddy, S., Ly, D. H. & Meller, A. Electronic barcoding of a viral gene at the single-molecule level. *Nano Lett* 12, 1722–1728 (2012).

5. Nir, I., Huttner, D. & Meller, A. Direct Sensing and Discrimination among Ubiquitin and Ubiquitin Chains Using Solid-State Nanopores. *Biophys. J.* 108, 2340–2349 (2015).

6. Squires, A., Atas, E. & Meller, A. Nanopore sensing of individual transcription factors bound to DNA. *Sci. Rep.* 5, 1–11 (2015).

452620 Measuring Macromolecular Properties in a Field-Free Single Molecule Trap

Abstract:

Mass and electrical charge are fundamental properties of biological macromolecules. While molecular mass has long been determined with atomic precision, we have achieved for the first time a high-precision (< 1 e), absolute measurement of the electrical charge of molecules such as nucleic acid oligomers, and globular and intrinsically disordered proteins in solution. The method is based on parallel, external field-free trapping of nanometer-scale entities in the fluid phase (1,2), requires one molecule, is rapid, and also facilitates an estimate of a dielectric coefficient of the molecular interior. We have further demonstrated direct detection of chemical modifications and single point mutations in proteins. Since net macromolecular charge strongly depends on 3D conformation, high precision electrometry will serve as a new approach to probe molecular structure, dynamics and interactions in solution, down to the level of the single molecule and in real time. The ability to experimentally link electrical charge and molecular structure will likely open up an unexplored physical dimension in biology.

References:

(1) Krishnan, M., Mojarad, N., Kukura, P. & Sandoghdar, V. Geometry-induced electrostatic trapping of nanometric objects in a fluid.
Nature 467, 692-695 (2010).
(2) Mojarad, N. & Krishnan, M. Measuring the size and charge of single nanoscale

objects in solution using an electrostatic fluidic trap. Nature Nanotechnology 7, 448-452 (2012).

452618 Nanobiodevice-Based Electrophoretic Separations of Single Biomolecule, Exosome, and Cell for Medical Innovations

Abstract:

Nanobiodevice is a piece of contrivance, equipment, machine, or component, which is created by the overlapping multidisciplinary activities associated with nanotechnology and biotechnology, intended for biological, medical, and clinical purposes. In this lecture, I will describe the development of nanobiodevices for biomedical applications, including single cancer cell diagnosis for cancer metastasis, circulating tumor cell (CTC) detection by microfluidic devices, nanopillar devices for ultrafast analysis of genomic DNA and microRNA, nanopore devices for single DNA and microRNA sequencing, nanowire devices for exosome analysis, single-molecular epigenetic analysis, quantum switching *in vivo* imaging of iPS cells and stem cells, and quantum technology-based cancer theranostics. Immunopillar devices realized the fast and low invasive "from blood to analysis" type biomarker detection of cancer with fM detection sensitivity within 2 min. Additionally, nanopillar devices give us ultrafast electophoretic separation of DNA and microRNA within 60 μs and nanopillar-nanopore integrated nanobiodevice enables us ultarafast single molecular DNA sequencing. Nanowire devices coupled with super-resolution optical microscopy are extremely useful to analyze exosomes from cancer cells and exosomal microRNA analysis. Nanowire devices are applicable to ultrafast and wide range analysis of DNA Molecules by rigid network structure of solid nanowires. Quantum dots are applied to develop quantum-biodevices for single cancer cell diagnosis, single molecular epigenetic analysis, quantum switching *in vivo* imaging for iPS cell (induced pluripotent stem cells) based regenerative medicine, and theranostic devices for cancer diagnosis/therapy.

454292 From Electrokinetics to Microfluidics and Back

Abstract:

Electrokinetics of electrolytes is a fascinating field, and probably one of the areas of physics in which numerous discoveries and surprises remain ahead of us, in spite of its rooting in the century-old Maxwell equations.

In this conference, I shall present some of my personal surprising encounters with this complexity,Â and how it brought me towards microfluidics. Iâ€™ll also present some examples of other fascinating microfluidic applications also rooted in the Maxwell equations, but rather on the magnetic side. They concern, in particular, to the use of magnetic microparticles for various analytical applications, focusing on attempts to bring this technology from the research laboratory to clinical applications. Finally, Iâ€™ll provide a personal view of some future challenges of electrokinetic issues in the microfluidic context. This will include recent attempts in our group, to use the electrokinetic instabilities often detrimental to conventional electrophoresis, as a tool for sensitive DNA detection. This work was supported in part by European projects NADINE LOVE-FOOD, ERC Cello, and French ANR project DIGIDIAG

AUTHOR INDEX

AUTHOR INDEX

AIChE
120 Wall Street, FL 23
New York, NY 10005-4020

ISBN 978-1-5108-3419-4

9 781510 834194

CURRAN ASSOCIATES INC.
proceedings.com